NUTRITION AND DISEASES MANAGEMENT OF POULTRY

NUTRITION AND DISEASES MANAGEMENT OF POULTRY

Ganesh Yadav

RANDOM

RANDOM PUBLICATIONS

NEW DELHI - 110 002 (INDIA)

Nutrition and Diseases Management of Poultry

ISBN 978-93-51117-01-8

Published in 2015 in India by

RANDOM PUBLICATIONS

4376-A/4B, Gali Murari Lal, Ansari Road
New Delhi-110 002
Phone: +9111-43580356, 23289044
E-mail: randomexports@gmail.com; sales@randompublications.com; info@randompublications.com
Reprinted 2019

Type Setting by: Friends Media, Delhi-110089
Digitally Printed at: Replika Press Pvt. Ltd.

Preface

Worldwide, more chickens are kept than any other type of poultry, with over 50 billion birds being raised each year as a source of meat and eggs. Traditionally, such birds would have been kept extensively in small flocks, foraging during the day and housed at night. However, rising world populations and urbanization have led to the bulk of production being in larger, more intensive specialist units. These are often situated close to where the feed is grown or near to where the meat is needed, and result in cheap, safe food being made available for urban communities. Profitability of production depends very much on the price of feed, which has been rising. High feed costs could limit further development of poultry production.

Nutritious feed, access to fresh, clean water, and adequate housing are important to the health of flocks. Good management and sanitation practices are essential as well. Proper ventilation in the brooder and coop will reduce moisture and disease organisms. Caked or wet litter should be removed as soon as it forms to keep the house clean and dry. As farming became more specialized, many farms kept flocks too large to be fed in this way, and nutritionally complete poultry feed was developed. Modern feeds for poultry consists largely of grain, protein supplements such as soybean oil meal, mineral supplements, and vitamin supplements. The quantity of feed, and the nutritional requirements of the feed, depend on the weight and age of the poultry, their rate of growth, their rate of egg production, the weather, and the amount of nutrition the poultry obtain from foraging. This results in a wide variety of feed formulations. The substitution of less expensive local ingredients introduces additional variations. Healthy poultry require a sufficient amount of protein and carbohydrates, along with the necessary vitamins, dietary minerals, and an adequate supply of water. Certain diets also require the use of grit, tiny rocks such as pieces of granite, in the feed. Grit aids in digestion by grinding food as it passes through the gizzard. Grit is not needed if commercial feed is used. The feed must remain clean and dry; contaminated feed can infect poultry. Damp feed encourages fungal growth. Mycotoxin poisoning, as an example, is "one of the most common and certainly most under-reported causes of toxicoses in poultry". Diseases can be avoided with proper maintenance of the feed and feeder. A feeder is the device that supplies the feed to the poultry. For privately raised chickens, or chickens as pets, feed can be delivered through jar, trough or tube feeders. The use of poultry feed can also be supplemented with food found through foraging. In industrial agriculture, machinery is used to automate the feeding process, reducing the cost and increasing the scale of farming. For commercial poultry farming, feed serves as the largest cost of the operation. Diseases caused by infection with a living microorganism such as bacteria, virus, mycoplasma, parasite, etc. are infectious diseases. Most infectious diseases are also contagious, that is, they spread from one chicken to another but a few, like Staphylococcus infection, and aspergillosis, are not. Sanitation is used to reduce the numbers of disease organisms which the chicken contacts to the level where they will no longer cause disease and to provide a clean, healthy environment. This can be done by cleaning and disinfecting, by adequate ventilation to reduce the number of organisms in the air and by reducing contact with other chickens by keeping them in cages. Some organisms are widespread in the environment and it is difficult to prevent birds from contacting them. Some organisms are found primarily or only in chickens (sick birds or healthy carriers) and live outside the bird's body for only a short time (mycoplasma). Others may live for days, weeks, or months in the environment, depending on the organism and the conditions. Workers must be trained to carry out the sanitation and isolation procedures. They must also understand why the rules are set up and why disease prevention is important. Medicine is often added to the feed at the feed plant. Most preventive medicine is used this way. In a disease outbreak medicine can be added to the water until medicated feed is available.

This essential information resource book reviews all aspects of poultry production, including layer production for eggs and broilers for meat. All people related to poultry business will find this book useful in their day-to-day practice.

I thank all members of my team who have helped in the preparation of the book. My special thanks go to "Random Publications" who have published the book.

— Ganesh Yadav

Contents

Chapter 1

Introduction

Nutritional Requirements of Poultry

Poultry require carbohydrates, fats, protein, vitamins, minerals, and water. Most nutrients provide both energy (as carbohydrate or fat) and protein. Feedstuffs are usually classed as being primarily a source of energy or protein. Fats and oils do not provide protein. Energy and protein are more efficient if they are available in the proper ratio. Excess protein is used as energy with increased waste excretion. Growing chickens should have 16-24% protein; growing turkeys, 24-28% protein; for egg production 15-17% protein; for maintenance (no growth or production) 10-12% protein. The highest requirement is in the first 2 to 3 weeks and is higher in young leghorns than broilers.

The protein level can be lower if the essential amino acids are all present at the proper level. The vitamin and mineral levels are higher for growing birds and for egg production. More rapid growth requires more added vitamins and minerals than slow growth. In slow growing birds more of some of these essential vitamins and minerals are available from the feed ingredients because the amount of feed required for basal metabolism is greater.

Some feed and water is required just for survival. The BMR varies directly with the weight of the bird. If the feed energy required to supply the basic energy need is not available, the bird will lose weight, get thin and die. Some protein, fat, vitamins and minerals are also required for basic survival. Nutrients available above the basic level can be used for growth in young birds and egg production in adults. Excess energy is converted to fat and stored in the body. It may be used when energy from feed is lacking as in the hump in camels and tail fat in fat-tailed sheep.

The rate of growth in young birds or egg production in adults is controlled by:

a) the genetic potential (commercial broilers compared to layers)
b) the amount of energy and protein (and protein quality) available above the level for the BMR. The feed required to maintain BMR does not add to growth or production. The maintenance requirement for a slow-growing broiler is the same as a fast-growing broiler of the same weight. The feed maintenance requirement for a hen laying 2 eggs a week is the same as a hen of the same weight laying 6 eggs a week. Sufficient vitamins and minerals must also be available.

If the temperature is below the birds comfort level, additional energy is required to provide internal heat by increased metabolism to maintain body temperature. Activity also increases metabolism and birds that are allowed to run outside or that scavenge for their feed have a higher requirement for energy.

Chickens that get all their nutrients from scavenging may eat an excess of protein, if insects, worms, larvae, etc. are available. They might benefit most from supplemental feeding of energy in the form of carbohydrate (cereal grains, etc.). Fenced or backyard poultry fed household or garden waste may lack both energy and protein for good growth or egg production.

Laying hens should have calcium available free-choice, even if calcium is being added to feed. Chickens that eat whole seeds, grains, vegetable material or fibre must have insoluble grit (granite grit) or small stones in their gizzard to grind the hard or fibre material. Birds that forage pick up their own grit. Poultry kept indoors must have grit supplied. Small chicks require small stones 2 to 4 mm. Hens .5 to 1.5 cm. Limestone particles are not satisfactory as grit.

The amino acids that growing poultry require are: arganine, cystine, lysine, methionine, tryptophan, histidine, isoleucine, leucine, phenylalanine, threonine, tyrosine, valine.

The following vitamins should be added to a prepared feed: vitamins A, D3, B12, riboflavin, pantothenic acid, niacin, folicin (folic acid), thiamine and biotin.

Commercial Nutrient Sources

In countries where complete feeds are prepared by feed manufacturers, cereal grains and oilseed meal are the major ingredients in the feed. Depending on cost, availability and age or type of poultry or water fowl, some of the following products are also frequently used: by-products of cereal grains processed for human use, animal processing waste, fishmeal, waste oil from restaurants (restaurant grease), yeast, alfalfa meal, distillery or milk factory products, etc. Vitamins and mineral are also added in the correct proportions. The ingredients are ground (if required), mixed and may be pelleted.

Ingredients most frequently used:

- for energy: maize (corn) up to 65%
- sorghum (milo) up to 45%
- wheat up to 25%; with enzyme up to 45%
- wheat by-products (bran, shorts, screenings) up to 15%
- rice up to 15%, rice by-products (bran, polishings) up to 15%
- barley up to 15%; with enzyme up to 35%
- molasses, up to 5% after 2 weeks
- for protein: soybean meal up to 30%
- soybeans up to 15% (heated to remove antinutrients)
- canola meal or whole seed up to 10%
- corn gluten up to 15%
- peas, lupin, flax up to 10%; flax (linseed) meal - 15 % (20% in layers)
- safflower meal, sunflower meal up to 10%
- meat meal, fish meal up to 10%
- blood meal, feather meal up to 2%
- Fats and Oils: tallow, lard, coconut oil, palm oil, after 3 weeks up to 5%
- poultry fat, fish oil, restaurant grease up to 3%, after 3 weeks 6%

Minerals: Calcium can be added as limestone, as prepared products such as dicalcium phosphate or as oyster or other marine shell. Growing birds require about 0.9% calcium in the diet. Hens in lay require 3 to 4% calcium in the diet so that calcium is available for egg shell production.

Phosphorus can be added from manufactured products such as dicalcium or monocalcium phosphate. Phytate phosphorus (an organic phosphorus) is present in small but varying amounts in plants feed ingredients. This phytate phosphorus is only 25-50% available unless phytase enzyme is added to the ration. Growing birds require about 0.4% phosphorus, adults about 0.3%. Calcium and phosphorus levels must be in balance since high levels of one cause deficiency of the other.

Sodium and chlorine are usually added as salt (NaCl) at about 0.25% of the ration (2.5 kg NaCl/1000 kg).

Magnesium and potassium are usually present in adequate amounts in the feed ingredients. Some ingredients such as soymeal and molasses are high in potassium. Minerals may be present in high or moderate levels in animal by-products. If animal by-products are being fed the minerals present in this feed should be used in the calculation of the amount to be added. High levels of sodium are toxic to birds. Saline water is also toxic for birds.

Trace minerals and vitamins are required in small quantities. They are not usually added as separate ingredients. They are usually bought as a prepared package or a premix. A specified amount of this premix can then be added to each tonne (1000 kg) of feed. To make sure of an even distribution in the final feed the vitamin-mineral premix may be mixed with 25-50 kg of cornmeal or other large particle ingredient that will help ensure an even mix. This is then added to the larger mixer after the cereal and protein portions have been put in. The calcium, phosphorus and salt should never be added to the vitamin premix. The mixer should be grounded.

Alternative Feeds

There are many vegetable and animal products that can be used as feed ingredients for poultry. Poultry cannot digest fibre found in roughage, seeds, grains and fruits. Products that are high in fibre can make up only a small percent of the feed. Since many plant products also contain antinutrients the amount of these feed stuffs that can be fed is also limited. Products that contain a high percent of water (kitchen and garden waste, weeds, tree and plant leaves, fruit, worms, larvae, fresh fish and animal waste, water animals and insects, milk products, etc.) can be given free-choice. What is not eaten must be removed so the chickens will not eat mouldy or spoiled material. Most feeds can be sun-dried to reduce the water content. They can then be ground into meal to be mixed with other feed, or cut up into chips that the birds can eat.

Many fresh (undried) foods such as garden and kitchen waste, fruit, plants and vines can be fed free choice. The best way to preserve foodstuffs is by sun-drying. It is important to avoid spoilage. Some types of foods can be preserved by anaerobic (no air) fermentation (making silage) where the fresh food ferments and becomes acid. The acidity prevents the growth of spoilage organisms. With plant material this may be done in a large plastic bag kept in the shade. Milk may be preserved for short periods by allowing it to sour (go acid) or by adding acid material.

Seeds and grains that contain antinutrients and toxins must be used with care since levels of antinutrients and toxins vary. Several varieties of raw or unprocessed peas and bean should not be used together (don't use 5% of each) because they contain similar antinutrients and toxins.

Proteins are fed for the nitrogen and amino acid content. Some amino acids are essential and must be present in the feed or added as synthetic amino acids. Poultry cannot make them from other amino acids or nitrogen. Feeding a variety of protein sources may supply more of the essential amino acids.

Poultry require both sodium and chlorine. There is some sodium and chlorine in feedstuffs, but both must be added to the feed, usually as salt. The recommended level is 0.25 to 0.3% added salt (2.5 to 3 kg per tonne). If feed ingredients or drinking water contains salt, the amount of added salt must be reduced. Poultry can be poisoned by the sodium in salt and young chickens are very susceptible to high sodium. Chicks are frequently poisoned by salt in the drinking water. Water for young chicks should not contain salt and many chicks die from salt poisoning at levels below 0.1% (0.9% is the level found in animal tissue). Chicks can be poisoned by eating salted fish and particularly by the brine used to preserve fish.

Young chicks can also be poisoned by high calcium levels in the feed. Feed for layers contains 3 to 4% calcium. That level can cause kidney disease in young chicks.

The Importance of Water

Water is an essential nutrient. All body functions and processes require water. Water is the material in which all other nutrients are carried in solution. The tissues and cells in the body are made up mainly of water. Water is also present between cells so that material can be moved between cells. Dry feed contains only 10-15% water and poultry need about twice as much water as feed (2 gms of water for each gm of dried feed).

Water should be clean and free from chemicals and minerals. It should not contain harmful parasites or bacteria. Water should be easily available free-choice. If water consumption is restricted, growing chickens will grow more slowly. Broilers that get only 2/3 or 1/2 of the amount of water they want will eat less feed and grow at only 3/4 of the expected rate. In cool climates water restriction is used to slow growth in breeders or to reduce the ascites syndrome. If water is not freely and easily available adults will lay fewer eggs and may suffer from kidney disease (as do people that do not consume sufficient water). Layers that go without water for a day or two will stop laying and may take 2-3 weeks to recover. Water in a trough or cup must be deep enough to allow the bird to drink properly. A depth of 2 cm is adequate, unless poor beak trimming has resulted in long lower beaks. The drinkers must be easy to reach and housed poultry must not have to walk too far to get water; not more than 2 metres for broilers and 4 metres for adults.

In hot climates water availability is even more important. The amount of water required becomes much higher as the temperature rises above 25BC. Birds cool themselves by water evaporation from their respiratory system and lose water when they mouth breath. In hot climates cool water is better than warm water. For large groups of poultry the water pipes to the poultry sheds should be 40 to 60 cm underground to cool the water as it flows to the chickens. Above ground pipes and emergency tanks should be shaded from the sun. To keep the water in emergency tanks fresh it can be used for 3 to 5 hours in the early morning. Chickens will move to cooler parts of the pen when the temperature is high, particularly if there is a breeze or fans. It is important to have water easily available in these areas because the birds may not go back to hot areas to drink.

Nipple drinkers help conserve water and avoid spillage, but chickens drink less water from nipples. Nipple drinkers require careful management. The pipes must be level, at the correct height (at eye level for chicks, just above eye level for older birds) and the water pressure must not be too high for the birds to operate the nipple. The water column that controls pressure should be about the same height as the chickens using the system (6 cm for a 6 cm chick).

Newly hatched chicks should have water before or at the same time as they receive their first feed. It must not be in an open pan or trough that the chicks can get into. Wet chicks lose body heat quickly and may die. Water jugs inverted in pans with a narrow water source are satisfactory as are raised narrow troughs, bell-type drinkers and nipple drinkers. Water jugs, troughs, etc. must be cleaned at least once a day.

Water Problems

Water containing physical particles or dissolved mineral material may interfere with automatic drinkers. Particulate material and iron may be removed by a sand filter and settling tank but dissolved minerals and algae growth in the system are more difficult to control. Regular cleaning and flushing may be necessary. Copper sulphate solution may control fungi and algae growth, but it can be toxic for poultry. Poultry may be poisoned by minerals or chemicals in the water. The sodium in saline (NaCl) water is the most frequent problem. Young birds are very susceptible to sodium toxicity. Sodium above 500 ppm (0.05%) in the drinking water may cause death in young chickens and turkeys depending on the sodium level in the feed. The young birds usually die from oedema and ascites syndrome. Wet droppings or diarrhea would also occur. Salt in the feed can be reduced to avoid disease from salt in the water up to about 1000 ppm (0.1%) of sodium in the water. Saline water above 1000 ppm should not be used for broilers up to 21 days of age even if no salt is added to the feed. High levels of sodium will kill adult chickens by causing diarrhea and dehydration.

Sodium may be present in water as sodium sulphate (NaSO4) or sodium bicarbonate (NaHCO3). Sulphate may be present as magnesium sulphate (MgSO4) and cause diarrhea and death. All sources of sodium in the feed, whether added as salt or present in the feed ingredients, (particularly animal protein) and water are additive in causing sodium toxicity. If sodium is too high in the feed it will cause disease as it does in the water. Sodium, however, is an essential nutrient and poultry require some sodium to grow and produce eggs.

Nitrate in feed and water may be toxic at high levels but can reduce growth at 50 ppm in water. A variety of other naturally occurring minerals and chemicals may be present in water and may cause problems. Surface water may be contaminated by farm pesticides, fertilizer or by industrial chemicals.

Brooding and Growing Chicks and Poults

Hatched chicks should be lively, firm and healthy. They should be uniform in size and have a well-healed navel. They should be started as soon as possible after removal from the incubators. The sooner they are provided a warm area, feed and water, the better the rate of success. Chicks must not be chilled or overheated at any time. If chicks do not receive feed and water within 18 hours after hatching they should get water for 1-2 hours before they are given feed.

Systems of Brooding

All pens and equipment should be ready, tested and operated before the chicks arrive.

Natural Brooding: For small numbers, a broody hen can handle 14-15 chicks. Protection from predators and rain is important. Chicks should get some commercial ration or other feed for the first 3-4 weeks.

Sour milk is good for young chicks. A little care over the first 3-4 weeks will often double or triple the survival rate.

Warm Room Brooding: In this system, the whole pen is held at a temperature of 30-32BC both day and night. The temperature is lowered about 1.5BC per week until the ambient temperature is reached, but should not go below 21BC until 6 to 8 weeks and not below 18BC until 10 to 15 weeks. Light should be over feed and water.

Circle Brooding: In this system, brooding areas of the pen under the heat reflector are heated to around 35BC by the use of reflectors over some form of artificial heat, such as gas heaters. Chicks are enclosed by cardboard chick guards, (not more than 400 chicks per circle) that allow them to move away from the heat source, but keep them close to feed and water. Light should attract chicks to the heat source. Temperature in the rest of the pen can be allowed to drop to 20 or 15BC.

Heat may be supplied by hot water or hot air with the heat source usually oil, gas, coal, wood or electricity. For small flocks of up to 75 chicks, a heat lamp may be all that is required. Temperature readings should be taken 5 cm above the litter (or at chick level). Make sure the thermometre is shaded from radiant heat.

With circle brooding, the feed and water should be placed at the edge of the heat reflectors, which is 60 to 70 cm above the floor. Comfortable chicks will be at the feeders and waterers and spread uniformly over the pen area. When chicks feel cold, they will crowd under the heat source or pile up. If the pen is too warm, the chicks move away from the heat with wings spread and/or are panting.

Uncomfortable chicks will be noisy:

(a) Floor space - Allow 1m2 per 75-150 chicks under the heat source and at least 1m2 per 25-50 chicks within the chick guard. Many producers use a chick guard of corrugated cardboard 30 cm in height placed 2-3 m back from the heat source for the first 5 to 7 days. The guard should be removed when the birds can fly over it. Broilers at 6 weeks require 1m2 per 9 to 12 birds.

    Leghorn chicks for egg production are often started in large (1 x 2m) or smaller growing cages with warm room brooding, or in hot climates with a heat source over each cage. Small growing cages may be about 50 x 50 cm with 24 chicks started in each. After 4 to 5 weeks, half the chicks are placed in another cage of equivalent size, to double the space per bird and split again to 6 per cage at 8 to 10 weeks (400 cm2 per bird).

(b) Feeders - Chicks should be given 5 (Leghorns) -8 (broilers) cm of feeding space. Feed is usually put in shallow boxes or egg flats on the floor or cage bottom for 4 to 6 days as well.

(c) Waterers - Good water is important. In large flocks, some form of automatic water system is usually installed. In a float controlled trough, 2 cm per bird is considered sufficient. One hanging fountain per 80 to 100 birds, one smaller cup per 50 birds, one nipple for 10-12 birds, will generally be adequate.

For small groups, three 4L or larger drinkers should be provided per 100 chicks, and more added as required. These should be cleaned and refilled daily. Feed and water should be within 1.5m of all chicks. Water should be clean and free from toxins and chemicals. Salt as NaCl or $NaS0_4$ is particularly dangerous in young chicks up to 21 days. Total sodium (Na^+) in water should not be above 300 ppm for broilers and 600 ppm for Leghorns or local strains.

Feeding

For best results, a commercial ration is recommended. If home grain or other feed is available, it may be fed starting at day 8 in broilers and day 15 in Leghorns. Replace 5% of commercial ration with whole grain or other feeds and increase by 5% a week in broilers (5% every 14 days in Leghorns) to a maximum of 50%, or a commercial concentrate may be diluted with whole or ground grain or alternate feed stuffs. When grain is fed, supply insoluble grit sprinkled on the feed at least once per week. Don't overfeed grit. Chickens raised for meat should be given all the feed they will eat (unless they are being raised at high altitude). Layers may have to be restricted to prevent them becoming too fat. Breeding flock must be monitored regularly and weighed weekly to make sure they are uniform and not overweight. Severe restriction may be required in broiler breeders.

It is advisable to include an anticoccidial drug in the feed of birds grown on the floor. This should be fed continuously throughout the growing period up to market in broilers and to 12 weeks in breeders or Leghorns. Coccidia vaccines are also available.

Cannibalism (Feather Pulling - Vent Picking, Toe Picking, Etc.)

Birds are cannibalistic by nature and when confined may start picking.

Chicks should be beak-trimmed at the hatchery, or at the farm using an electric trimmer. Broilers should not need trimming but watch for signs of picking and beak-trim if necessary.

Other steps to consider are:

- decreasing the light intensity;
- decreasing pen temperature;
- provide more space;
- use red light bulbs or place red plastic over windows;
- feed whole oats if the flock is over 4 weeks of age;
- add 1 g salt per L of drinking water if flock is over 6 weeks of age.

Once started, cannibalism is difficult to control.

Trouble Shooting

- Chicks should be well spread out and at the drinkers and feeders
- If noisy and crowding the walls or chick guard they may be too warm
- If noisy and crowding the heat source they may be too cool or drafty
- If crowded around feeders or drinkers, check supply and space per bird.
- If chicks are huddling or piling in small lots they may be too cold, too hot or sick.
- Litter condition
  - — too dry, may cause respiratory problems - spray litter to increase humidity.
  - — too wet, ammonia and coccidiosis or leg problems occur - check ventilation and bird floor space.
- Insufficient feeder or waterer space, disease, poor lighting, etc., will all affect uniformity.

Poultry Health Management for Commercial Poultry

Disease Prevention

Diseases caused by infection with a living microorganism such as bacteria, virus, mycoplasma, parasite, etc. are infectious diseases. Most infectious diseases are also contagious, that is, they spread from one chicken to another but a few, like Staphylococcus infection, and aspergillosis, are not.

Prevention by Sanitation

Sanitation is used to reduce the numbers of disease organisms which the chicken contacts to the level where they will no longer cause disease and to provide a clean, healthy environment. This can be done by cleaning and disinfecting, by adequate ventilation to reduce the number of organisms in the air and by reducing contact with other chickens by keeping them in cages.

Sanitation affects all levels of the birds' environment:

- *Building and Equipment:* Cages are intended to keep birds out of contact with their droppings, spilled feed and water, etc. that are a source of bacteria and bacterial growth. Fly control is important if this method is to be effective. On floor operations, an adequate amount of dry litter helps control bacterial growth.

  At cleanout, blow or spray dust from walls, ceiling, rafters, fans, etc. Wet down and remove litter, clean the building and equipment using detergent and water. Follow with a good disinfectant. A dirt flood cannot be disinfected. Litter should be piled or spread at least 100m from the poultry buildings (1 km is prefered).

- *Feed:* Between flocks, clean out feed in the troughs and bins both inside and outside the building and remove it from the building. It can be held to feed to chickens of a similar age or sent for reprocessing. Outside storage bins must be checked to make sure spoiled or moldy feed is not adhering to the inside and that the bin is empty. Feeders should be constructed and set so that feed is not spilled or billed out into the litter.

- *Water:* Open troughs are a source of contamination from the nasal and oral secretion and feces, etc. and must be cleaned regularly. Nipple drinkers are much cleaner but pressure must not be too high. Tremendous numbers of bacteria can build up in ponds and some wells, in water lines, and in drinking cups or troughs, so a periodic or constant water-sanitising programme and descaling may be necessary.

- *Air:* Clean, germ-free air is a very important part of a healthy environment. A good ventilation system to dilute and carry away dust and microorganisms and keep the pen dry is part of a sanitation programme.

- *The Caretaker:* Workers can carry infection to birds on their hands, clothes, boots, and equipment. They can spread germs from one group to another on the farm or bring germs in from outside the farm. If the same person is looking after more than one building, outer clothing and footwear should be changed between buildings and hands washed in disinfectant. If possible, separate workers should care for chicks, young birds and adults.

Good isolation requires shower-in and no contact with other chickens or other people who have chickens or work with chickens. Sanitation is a method of eliminating or reducing the number of disease causing organisms from contacting the birds. Although one germ may not cause disease and healthy birds may resist hundreds of germs, sanitation cannot be relied on to control serious contagious diseases. It must be used in conjunction with isolation and sometimes vaccination for problems such as Newcastle disease, avian influenza, infectious laryngotracheitis, fowl pox, etc. However, sanitation is important in maintaining the general health of a flock and is an important part of good flock management.

Prevention by Isolation

This method of disease control is simple. Stop the microorganisms that cause disease from contacting the chickens. When the microorganisms which cause a disease are eliminated from an area or country, the disease is said to be eradicated.

Whether or not a disease can be prevented by isolation depends on:

(a) Where the microorganisms that cause the disease live. Some organisms (like the ones that cause necrotic enteritis, staphylococcosis, coccidiosis, *E. coli*, etc.) are widespread in the environment and it is difficult to prevent birds from contacting them. Some organisms are found primarily or only in chickens (sick birds or healthy carriers) and live outside the bird's body for only a short time (mycoplasma). Others may live for days, weeks, or months in the environment, depending on the organism and the conditions (moisture, temperature, etc.).

- The way the disease organism is spread. Most disease germs are spread by direct contact, that is bird to bird. If the disease germs get on poultry fluff or small, air-borne particles, they can be carried by the wind, but this usually only over short distances. Organisms are frequently spread by mechanical transfer by carrier objects such as people, eggs, egg cases, trucks, feed, water, rats, dogs, insects, etc. Two or more separate objects may be involved in the spread of organisms from one flock to another (chickens to eggs, to egg cases, to people, back to chickens). Some animals, and certainly wild birds, can become infected with a disease and spread the germs. Insects are a necessary part of the spread of some diseases (leukocytozoon, plasmodium, etc.). Some disease organisms are spread through the egg from the adult to the chick (vertical spread). Some disease germs are spread venereally or by artificial insemination in semen (mycoplasma).

Diseases that are easier to control by isolation are those where the causative agent survives outside the body for only a short time. If the agent is egg-transmitted, the breeder flock must also be free of the disease. With some diseases a regular monitoring programme may be required to insure that the flock remains negative.

Isolation is not only the cheapest but the best way to control many contagious diseases and good isolation is equivalent to a quarantine.

Important Features of an Isolation Programme

Workers must be trained to carry out the sanitation and isolation procedures. They must also understand why the rules are set up and why disease prevention is important.

- Have only one age group on the farm (an all in, all out programme). Buildings over 100 metres apart can be treated as separate units if proper isolation and sanitation procedures are followed.
- Obtain chicks or replacements from a disease-free, adequately isolated, single source or raise replacements in a different area with separate caretakers.
- Have no neighbouring poultry buildings or free ranging chickens within 300 metres.
- Clean and sanitise buildings and equipment between crops. (wet down litter before removal to protect neighbouring poultry and do not store or spread litter near poultry buildings).
- Screen buildings against wild birds and keep out rats, cats, and dogs, and control insects. Remove dead birds from the pens at least twice a day and dispose of sick and dead birds at least 100m from the poultry buildings. Make sure dogs, cats and wild birds or animals cannot drag or carry dead chickens onto the farm.
- Limit the movement of workers from one building to another.
- Bring in only new or sterilised egg cases and flats.
- Make sure employees do not keep poultry or pet birds or come in contact with free-range chickens or their droppings and do not visit other poultry farms.
- Keep out visitors (particularly those who may visit other poultry farms) and provide boots and protective clothing for persons entering the poultry area.

- Disinfect necessary vehicles (feed trucks, etc.) and restrict them to the loading and unloading areas which should not be near the building entrance. Keep the driver in the truck or provide boots and coveralls.
- Make sure poultry service crews disinfect equipment, shower, and change clothing before entering the poultry area (except at cleanout).
- Shower and change clothing after taking chickens to market or meeting with poultry workers from other farms.

Prevention by Vaccination

Poultry have a good immune response to many diseases and to vaccination. They also pass immunity to offspring through the egg. Breeders require a special vaccination programme.

Marek's Disease (MD): There are several types of vaccine. All are live and given by injection. If vaccine is available, all chicks should receive MD vaccine immediately after hatching. Frequently two types of vaccine are given together. A second vaccination at a later date is not required. Chicks must be kept in a clean pen, away from other chickens for three weeks.

Newcastle Disease (ND): There is only one serotype of NDV so proper vaccination protects against all pathotypes of virus. Maternal antibody is provided for several days by immune hens. In countries where velogenic virus is endemic, chicks should be vaccinated before day 7 at day 21, 35 and 48 using more immunogenic (or virulent) vaccine (LaSota) for the 3rd and 4th vaccinations. Vaccinations should be repeated at 45 to 60 day intervals in places where ND is endemic. Properly administered live spray vaccine is best in adults. Eye drop administration is best for chicks. Either eye drop or spray can be used after 8 weeks. Vaccine in feed is available for village flocks. Killed vaccine may be given by injection.

Infectious Bronchitis (IB): There are many serotypes of IBV. Although there is some cross protection, new serotypes appear regularly that can evade the immunity of some vaccines. Commercial layers should be vaccinated as recommended locally. Broilers may also require protection. Live vaccines given in water or by spray are best in growing birds. Killed vaccines usually in oil may be used for adults. Some serotypes of IBV cause kidney disease.

Infectious Bursal Disease (IBD): There is one major serotype of IBDV but there are many pathotypes. As with ND, mild strains of vaccines will not protect against highly pathogenic strains of virus. Maternal antibody is provided for 7 to 14 days by immune hens. Where pathogenic strains of virus are present, moderate strength vaccine should be given at days 8, 14, and 28. If maternal antibody is not uniform, mild vaccine can be given at day 1. For maternal immunity killed vaccine in oil (often with ND & IB) may be given twice at 16 & 18 weeks and may be repeated at 35 to 45 weeks if necessary.

Avian Encephalomyelitis (AE): Epidemic tremors. Only breeders must be vaccinated but egg layers should be done as well to prevent a drop in egg production. Vaccination is done at about 12 weeks by water or wing web stab (live vaccine).

Viral Arthritis (VA): The arthritic form of reovirus infection is rare in many countries. Where the disease occurs broiler breeders should vaccinate twice to provide protection to chicks (live or killed vaccine).

Mycoplasma Gallisepticum (MG): Live MG vaccine is most effective if it is given to MG negative chickens before they become infected by the field strain. It provides less protection if given before 4 weeks. Other programmes are available for breeders.

Infectious Laryngotracheitis (ILT): ILT can be prevented by good isolation and sanitation procedures. The vaccine virus is live and may revert to pathogenic virus and is shed for life by vaccinated birds (as is field virus by survivors of outbreaks). In endemic areas breeders and egg type chickens should be eye drop vaccinated when moved to the laying house. Vaccinate 2 weeks before infection is likely to occur if growing birds are at risk. In some places broilers are vaccinated in the drinking water at 3 weeks.

Fowl pox: In endemic areas pox may be spread by mosquitos and cannot be prevented by isolation. Live pigeon pox (or other types) vaccine may be given by wing web stab at 8 to 12 weeks. Turkeys may also require protection.

Egg Drop Syndrome (EDS, Adenovirus Group III): In countries where EDS occurs layer and breeder flocks may be vaccinated at about 16 weeks with killed oil-adjuvant bacterin.

Adenovirus Group I (Inclusion body hepatitis, IBH): There are 12 or more serotypes of adenovirus Group I. Subtype 4 and 8 may cause disease without previous immunosuppression. Live vaccines may have to be prepared to prevent these infections.

Fowl Cholera: On farms or areas where fowl cholera occurs, birds may be vaccinated with killed or live vaccine. There are many strains of Pasteurella multocida and autogenous bacterins may be required for protection.

Infectious Coryza: As with fowl cholera. Live vaccine is not available.

Coccidiosis: Chickens or turkeys raised on litter or outdoors should be protected by medication until they develop immunity or by oral vaccination at day 1.

Turkeys may be vaccinated against Adenovirus Group II (which causes hemorrhagic enteritis (HE) with live vaccine and against erysipelas with bacterin.

Prevention by Medication

- Preventive medication Some diseases such as coccidiosis, necrotic enteritis and enterohepatitis can be prevented by medication. Preventive medication is most useful when protection is only required for a limited time as in broiler chickens or when immunity does not develop such as in necrotic enteritis.

  Coccidiosis: Anticoccidial medication is very widely used in broilers and is generally used until the birds are ready for processing. Resistant strains of coccidia may develop and anticoccidials may have to be switched regularly. Immunity may or may not develop and anticoccidials that allow immunity are best.

  Necrotic enteritis (NE) (Clostridium perfringens type A): Low level antibiotic in the feed is used to prevent NE and in Quail ulcerative enteritis (UE) (Clostridium colinum). The bacteria must be sensitive to the drugs being used and if predisposing factors such as coccidia infection are present they must also be corrected.

- *Therapeutic Medication:* Therapeutic medication can be considered preventive when it is used to control the spread of serious infectious diseases such as coryza or cholera. Treatment of coryza should not start until 1/3 of the flock is infected or the disease may recur following treatment.

  Some bacteria quickly develop resistance to medicine and different drugs may have to be used for control. Virus infections do not usually respond to medicine, although the medicine may prevent bacterial infection secondary to disease caused by virus.

  Medicines given by injection should not be given into the abdomen or leg. They can be given under the skin of the back or into the muscle of the breast.

Medicine given in the drinking water can be poured into the drinkers. In an automatic system they can be mixed in a large container and run into the system by gravity or a pump. Medicine can be added to a pressure system with a proportioner. Birds drink more water in hot weather. The level of medicine must match daily consumption and should be reduced in hot weather. It could be given for just 8 to 16 hours a day.

Medicine is often added to the feed at the feed plant. Most preventive medicine is used this way. In a disease outbreak medicine can be added to the water until medicated feed is available. Growing birds drink about 3 times as much water compared to the amount of feed they eat. Adults about 2 times as much (by weight).

Diagnosis of Poultry Disease

History. A good history will often provide clues that will help solve a problem. Get information on the type of bird, age, feed and water source and consumption rate, growth, production, morbidity and mortality, the description of the case, previous problems, vaccination programme, medicine being used, etc.

The problems may relate to management, environmental factors, and stress rather than to infection so examine the yard and housing conditions. Is the ventilation adequate? Are ammonia fumes a problem? Is it too hot or too cold? Is the litter wet or is it too dry and dusty? Is the pen too light or too dark? Are there sufficient hours of light for best production? Is the nest area darkened? Are the roosts too high? Do the birds appear comfortable? Chickens can talk and the sounds they make can indicate comfort, hunger, pain, panic, or disease. Examination of Live Birds. Check the general appearance of the individual or group and try to determine which organ or system is involved in the illness. Note any signs or lesions that might point to a diagnosis, such as small size with poor feathering that suggests infectious stunting (malabsorption syndrome). If the birds show lameness or paralysis, is the lesion in the nervous system, bones, joints, muscles or skin? Some conditions, particularly those affecting locomotion, are easier to diagnose in live birds. Botulism which produces neck paralysis in chickens (leg and wing paralysis are more obvious in turkeys, ducks and pheasants) is an example.

Examine the skin of the head, body and legs for lice and mites, injury (particularly cannibalism), blood, mottling, swellings, anemia, cyanosis, or dermatitis. Listen for unusual breathing sounds (snicking, gurgling) and look for gasping or head-shaking that might indicate respiratory distress. Mouth-breathing (panting) is normal in chickens in hot weather. Exudate from nostrils and eyes and dirty feathers also suggest respiratory infection, or if just the eye, ammonia burn, ILTor eyeworm. Examine the droppings for evidence of diarrhea or other abnormalities. Take a blood sample for hematology or serology if indicated.

Necropsy. If a postmortem examination is to be carried out, birds that are representative of the problem in the flock must be selected. If birds have died, both sick and dead birds should be opened. Cull birds will not provide the answer. If the problem is a drop in production, try to find birds that look like they have recently stopped laying. It is important to do both an external and internal examination and to follow a specific routine to avoid missing important lesions.

- CRD - chronic respiratory disease
- IB(V) - infectious bronchitis (virus)
- ND - Newcastle disease
- AI - avian influenza
- CA(V) - chicken anemia (virus)
- EDS - egg drop syndrome
- IBH - inclusion body hepatitis
- ILT - infectious laryngotracheitis
- IBD - infectious bursal disease.

Live birds may be killed by cervical dislocation except when anemia or respiratory disease is suspected.

Examine the head, including the eyes, ears, nostrils, comb, wattles, mouth and beak.

Infected eyes and conjunctivitis are seen in CRD, sinusitis, Bordetella and Orthobacterium infection, infectious coryza and in ILT, IB and ND.

Keratoconjunctivitis with central corneal ulcers suggest ammonia burn. Swollen sinuses are seen in infectious coryza (Parahaemophilus infection) in chickens, sinusitis (mycoplasma), cryptosporidia and Bordetella infection in turkeys and pheasants. They may also be seen as the result of other infections such as influenza.

Chronic fowl cholera causes swollen wattles and sinuses in adult chickens. Fowl pox causes scabs on the comb, eyelid, and wattle, but must be differentiated from injury. Pox also occurs in turkey, pigeons, doves, canaries and other birds. If the bird has had its beak trimmed, check for proper healing, overgrowth of the lower beak or over-trimming (cut too short).

Injury on the head, neck or breast or back may indicate predators. Dermatitis and scabby or crusty lesions around the mouth and eyes suggest vitamin B deficiency.

Cut across the upper beak and back into the sinuses; then through the mandible and down the esophagus into the crop.

White plaques in the mouth, esophagus, or crop may be caused by capillaria worms, yeast infection (candidiasis, moniliasis), or possibly trichomoniasis or vitamin A deficiency, but most frequently by fowl pox (wet form). White plaques beside the tongue in the mouth are common in hens and turkey breeders and may be caused by mycotoxin or low humidity. Vomitoxin may produce similar lesions in young chickens and turkeys.

If the crop is enlarged and full, it may be an impacted or a sour crop (pendulous crop). The problem may be caused by excess water intake, defects in the crop itself, partial blockage of the proventriculus or gizzard or Marek's disease. Necrosis of the crop in wild birds is caused by Salmonella infection.

Examine the soft palate and larynx and cut down the trachea.

Wet fowl pox lesions are seen on the roof of the mouth, pharynx and larynx.

Granulation, congestion and mucus in the trachea are seen in CRD, IB, coryza, ILT, Orthobacterium, Bordetella, and *E. coli* infection, or with dust and ammonia fumes.

Hemorrhage and blood clots may occur in ILT and in ND or coryza and may cause severe gasping.

Gapes (gape worms) in pheasants, quail, and other birds is caused by Syngamus infection which also causes gasping. Cyanthastoma cause similar infection in water fowl and tracheal flukes may be found in waterfowl. Tracheal and air sac mites occur in cage birds.

Check the abdomen for lice and mites, and the vent for injury and prolapse of the oviduct.

Cut down between the abdomen and the legs and dislocate the hip joints. Peel the skin off the abdomen and breast. Tightly adhering skin and dark tissues indicate dehydration. Remove the breast carefully by cutting through the abdominal muscles, ribs and coracoids.

At this point, examine the thoracic and abdominal organs, paying particular attention to the air sacs, lungs, and liver. Carefully raise the gizzard and intestines and examine the abdominal air sacs, peritoneum, spleen and the ovary in laying birds.

The air sacs are cloudy in respiratory disease such as *E. coli* infection, Mycoplasma, Aspergillus, IB or early CRD in chickens. If there is also fibrin on the liver and in the pericardial sac, suspect *E. coli* or Salmonella infection. These lesions in turkeys and other birds might also be due to Chlamydia, or in ducks Anatipestifer infection.

Pneumonia in turkeys is caused by fowl cholera (Pasteurella multocida infection), and the lungs are quite solid. Aspergillosis, ND, AI, E. coli and infection can also cause pneumonia.

Peritonitis in layers is usually "egg peritonitis" caused by *E. coli* infection from the oviduct, although acute fowl cholera also causes peritonitis.

White crystals on the heart, liver and other tissues and organs are uric acid crystals (visceral gout) and are caused by hyperuricemia from urate nephrosis secondary to water deprivation, urolithiasis or other kidney disease.

Large blood clots in the abdomen or hemorrhage and hematoma on the liver are the result of trauma or fatty liver hemorrhage syndrome in chickens, or ruptured aorta in turkeys.

Tumors in or on the organs may be Marek's disease, lymphoid leukosis, or other varieties of tumors. Multiple small tumors on the organs and peritoneum in adult hens are metastasis from a carcinoma of the oviduct. This may result in ascites.

Ascites may also result from heart or liver disease or from ingestion of some toxic material. Right ventricular failure occurs most frequently in meat-type chickens after 4 weeks. It occurs secondary to pulmonary hypertension causing right ventricular hypertrophy and valvular insufficiency.

Focal white lesions on the organs may be tumors or they may be tuberculosis or coli-granuloma or, if just on the liver, blackhead. Turkeys, pheasants, and peacocks are all quite susceptible to blackhead; chickens are less susceptible. There are prominent cecal cores in blackhead (Histomoniasis) and occasionally in salmonella infection or coccidiosis.

In chicks, turkeys and waterfowl, examine the lungs and air sac for yellow-white foci or plaques caused by brooder pneumonia (Aspergillosis). Gasping in young birds is a sign of tracheal or bronchial epithelial injury, or obstruction and can be caused by irritating fumes or by infection (often IB plus *E. coli* or Aspergillosis).

In young chicks, look for an infected navel and for yolk-sac infection or peritonitis (mushy chick, omphalitis) in which the abdomen is swollen, wet, and discoloured, and the yolk-sac is infected (due to *E. coli*, Salmonella, Staph., Proteus, etc.).

Young birds also die because they don't start to eat (starve-outs) or drink (dehydration). Deaths occur mainly at 3, 4, and 5 days. This may be a management problem (chilling, feed and water not available, etc.) or the chicks/poults may be weak or defective when hatched.

Broiler chickens that die suddenly from sudden death syndrome (dead in good condition) or from heat stroke, or suffocation (piling-up) have congested, edematous lungs, a full digestive tract and congested mottled breast muscle.

Examine the circulatory and immune systems, the heart, pericardial sac, blood, spleen, bursa, thymus, and lymphoid tissue of the thigh and intestine.

Birds that die from anemia are pale and the blood is watery. With CAV the thymus is small and bone marrow may be pale. Birds that bleed to death (pick-outs, ruptured fatty liver, acute cecal coccidiosis, ruptured aorta, hemorrhagic enteritis in turkeys, etc.) are also pale.

Sulfa poisoning produces anemia and widespread hemorrhage in the tissues. Chicken anemia virus (CAV) produces similar lesions and is the agent responsible for infectious anemia associated with IBH. Lead poisoning may also cause anemia.

To identify anemia from parasites in blood cells (Plasmodium or leukocytozoon in ducks, turkeys and chickens), blood from a live, sick bird must be examined.

IBD (Gumboro disease) is caused by a virus that damages the bursa causing illness in 2-4 week old chickens or in younger chickens destroys part of the immune system, making birds more susceptible to other infections.

Swelling, congestion and hemorrhage with or without focal necrosis in the spleen, liver and lymphoid tissue suggest septicemia (fowl cholera, fowl typhoid, streptococcosis, or erysipelas) or viremia (ND and duck virus enteritis also affect the lymphoid tissue in the intestine).

Marek's disease and lymphoid leukosis produce tumors in lymphoid tissue and organs except for the bursa which is mainly affected by lymphoid leukosis (occasional Marek's lesions may occur in the stroma of the bursa).

Skin leukosis is Marek's disease virus causing viral dermatitis in the feather follicles. At processing this can be confused with scabby hip or other causes of dermatitis. Marek's disease can also cause lymphoid neoplasia in the skin.

Cut through the proventriculus and remove the digestive tract and liver. Open the proventriculus, gizzard, and small and large intestines to the cloaca. Check the cloaca carefully for evidence of picking injury.

If hens are not properly beak-trimmed or are too fat, mortality from "pick-out" is common. The whole intestine may be picked out through the cloaca. Prolapse of the vagina (and cloaca) (blow-out) may occur from excess fat or straining, secondary to injury or inflammation. Injury is common in flocks that produce large eggs before they become mature.

Examine the digestive tract for lesions and the various kinds of enteritis (hemorrhagic, necrotic, ulcerative, etc.), parasites (tetrameres, roundworms, capillary worms, tapeworms, cecal worms, and coccidia), and gastrointestinal accidents. A large proventriculus in broilers is from lack of fibre in the diet resulting in poor development of the gizzard. A thickened proventriculus sometimes with ulcers or hemorrhage may be Marek's disease or infectious proventriculitis.

Green staining of the digestive tract is just bile and indicates that the bird is not eating. The liver and spleen may be small (if the bird is thin) and the gallbladder full. Salmonella pullorum (pullorum disease, bacillary white diarrhea) causes enteritis, diarrhea and death in chicks. It has been eradicated in many countries.

Check the ceca, intestine and liver for lesions of blackhead, TB, coccidiosis or tumor, the liver for other varieties of bacterial, viral or protozoal hepatitis, cholangiohepatitis, etc., and the pancreas for tumors. A large liver may be lymphoid leukosis or Marek's disease, or it may be caused by bacterial hepatitis (*E. coli*, campylobacter) or fowl typhoid (Salmonella gallinarum).

Hepatitis with necrosis and hemorrhage that looks like bacterial (vibrionic) hepatitis may be immune damage to veins (vasculitis) from amyloid.

A large, yellow liver may be normal fat storage in a laying bird (estrogen stimulation) but layers die from fatty liver hemorrhage syndrome. This occurs when the liver ruptures because it has become fragile due to excess fat and free radical damage from fat.

A yellow or hemorrhagic liver particularly with focal necrosis may be viral hepatitis (IBH) which is seen in broilers, pigeons, raptors, owls and psittacine (but IBH does not cause infectious anemia).

Examine the testes or ovary and open the oviduct. Shrinking ova indicate illness of 2-7 days' duration or one from which the bird may be recovering. Small, sac-like ova indicate that the bird has been out of lay for a week or more and may be in a molt.

Semi-solid (cooked) ova indicate bacterial infection (such as salmonella).

An impacted oviduct may be secondary to vent-picking, egg-material left in the oviduct, or the bird may be egg-bound. Infection (Mycoplasma, IB, *E. coli*) can cause salpingitis as well.

A large or small fluid-filled cyst in the right abdomen beside the cloaca is the cystic remnant of the right oviduct (may be normal, but increased by IBV).

A drop in production may be related to clinical or subclinical disease (EDS, IB, MG, AE, AI, ND, etc.) or management faults (lack of light, temperature change, lack of water) or nutritional problems, etc.

Deformed shells suggest management faults, superovulation, EDS, IB and soft shells (higher than 1-2%) calcium or vitamin D3 deficiency.

A normal-appearing dead bird with an egg in the shell gland, or just laid has likely died from acute hypocalcemia. These birds often have fragile bones (particulary the femur), lack of medullary bone and rib infolding (osteoporosis, osteopenia, osteomalacia).

Hard (fibrotic) or swollen testes indicate bacterial infection (salmonella).

Examine the kidneys and ureters.

Swollen, pale or white spotted kidneys indicate hyperuricemia from urate nephrosis and may be due to lack of water or other kidney disease. Swollen. pale kidneys are also seen in IBH, IBD and fatty liver and kidney disease. Ureters plugged with hard stony material (urolithiasis) indicates a previous low phosphorus diet.

Swollen kidneys and nephritis may be due to IBV (nephrotrophic strain) or *E. coli* infection and usually causes death from dehydration. Newcastle's disease causes lympholytic foci in the kidney in pigeons.

Examine the skin, integument, muscles, bones and joints.

Emaciation, along with small organs, suggests malnutrition, stunting syndrome, beak injury (poor trimming), peck order (psychological) problems, chronic disease (coccidiosis), bumblefoot or other lameness, or chronic poisoning (lead, insecticide, etc.).

Muscular degeneration due to vitamin E-selenium deficiency can cause lameness, particularly in ducks. Ionophore toxicity causes muscle damage and paralysis in turkeys. Granulomas in the breast muscle are usually a vaccine reaction.

Sarcosporidial cysts produce small, white lesions in the muscle of water fowl.

Examine bones and joints for abnormality and deformity. Angular bone (valgus-varus) deformity of the intertarsal joint is caused by lateral or medial bending of the tibio-tarsal and metatarsal bones and is a common problem in meat-type poultry. It has a variety of possible causes (nutritional, rapid growth, management, etc.). Tibial dyschondroplasia causes backward bending or fracture of the top of the tibia. Slowing growth in young birds will help prevent leg deformity.

Other types of hock and stifle lameness and ruptured tendons are frequent in heavy roaster and turkeys and may be mainly due to injury as the result of heavy weight and fast growth, but some respond to added selenium or B vitamins.

Check for poor bone-breaking strength (osteoporosis) and, in young birds for rubbery bones, soft beaks, and beaded ribs which indicate calcium or Vit. D3 deficiency (rickets). Cage layer fatigue because of fragile bones may be due to phosphorus deficiency. Calcium and Vit. D3 deficiency also cause fragile bones (osteoporosis) in adults, but the most common cause is continuous high production.

Infectious stunting syndrome (malabsorption syndrome, fragile bones, osteoporosis) in young broilers is caused by intestinal damage from viral infection in young chicks. The chickens are small, poorly feathered and there is poorly digested food in the lower intestine. Osteomyelitis also causes fragile bones.

Curly-toe paralysis in young birds may be riboflavin deficiency, but in older birds and turkeys may be genetic or due to lack of roosts.

Cracked feet and foot dermatitis may be pantothenic acid or biotin deficiency, but scaly leg is caused by mites.

Footpad dermatitis and hock lesion are often caused by poor litter conditions. Toe injury in young birds may be cannibalism or mechanical injury. Arthritis in the feet or hocks or other joints suggests infectious synovitis (Mycoplasma synoviae) or *E. coli* or Staph. infection often with osteomyelitis.

Infection in the wing joints in pigeons is usually due to salmonella.

In broilers, roaster and broiler breeders viral arthritis (reovirus infection) may cause lameness or ruptured tendons. Ruptured tendons are usually caused by rapid growth and large body size (heavy weight) and there is usually thickening of the tendon above the hock.

If growing birds are lame, and there is no evidence of infection or rickets, cut into the proximal tibia and look for necrosis caused by osteomyelitis or dyschondroplasia and split the spine at T4 to look for spondylolisthesis (kinky-back), a plug of cartilage impinging on the cord or consider Marek's disease.

Necrotic dermatitis is caused either by staph. or clostridium infection and is associated with immunosuppression (usually by CAV or IBD). Scabby-hip is usually from overcrowding, poor litter conditions, poor feathering or scratches. Dermatitis or granulation in the neck may be a vaccine reaction, or contaminated vaccine.

Disturbances of the nervous system may cause incoordination, staggering, paralysis, walking backwards (with wings flapping for balance), tremors, stargasing, and other odd behaviour.

In case of lameness, paralysis, or incoordination, examine the sciatic nerves, spinal cord, and brain. Histologic examination will be required for diagnosis.

ND may produce CNS disturbances in pigeons as well as nervous, respiratory, intestinal and reproductive lesions in chickens, pheasants, turkeys, and wild and cage birds of all ages.

Range paralysis is a form of Marek's disease affecting the peripheral and central nervous system causing lameness, incoordination and paralysis. Marek's can also cause botulism-like paralysis and transient paralysis.

Vitamin E deficiency (avian encephalomalacia) causes soft, dark areas of hemorrhage in the cerebellum which may be visible grossly. B vitamin deficiency (thiamine and other) may also cause nervous signs.

Avian encephalomyelitis (AE) (epidemic tremor) affects birds up to 3-4 weeks old from non-immune parents.

Arsenilic acid and other feed additives and toxins (botulism, lead) may cause CNS disturbances, while others like the ionophores and coffee weed seeds (Cassia) cause nerve or muscle damage. Bacterial infection (pasteurella, pseudomonas, salmonella, staph. etc.) and fungi (aspergillus, etc.) also cause meningo-encephalitis, occasionally in outbreak proportions.

Hatching Eggs and Incubation

The Breeder Flock

Healthy chicks come from good breeders. Breeder nutrition, uniformity and the vaccination programme are critical.

Breeder Flock Health, Age and Nutrition

A healthy breeder flock, females with a good rate of lay mated to vigorous males, will usually produce highly fertile eggs that hatch well under good incubation practices. Hatching eggs from new breeder flocks (just starting to lay) do not hatch as well as eggs from those that have been laying for three to four months. Hatchability of eggs from old breeders nearing the end of their laying period also declines.

A hatching ration should be fed to breeders for about three weeks prior to saving eggs for incubation. The extra vitamins and minerals that it contains, as opposed to a laying ration, are essential for good hatchability. Sometimes, poor hatching results occur when a lower priced laying ration is used. The nutritional requirement of the breeder hen must be adequate for optimum hatching and good chick quality.

Clean Eggs are Important

Breeder house sanitation, nest box sanitation and hatching egg care are important to the production of healthy chicks. Nest material must be clean and dry and free of contamination. It should be cleaned daily (remove dirty material and droppings) and replenished or changed weekly. Para-formaldehyde pellets (20-25 g per nest, per month) will improve nest-box sanitation.

Breeder farms require a clean area where eggs can be cleaned, selected for quality and placed large end up on hatching trays. The temperature of this room should not be less than 21BC. If eggs are stored on the farm, temperature and humidity must be controlled.

Selection and Care of Hatching Eggs

Set only clean eggs. If eggs are soiled, it is preferable to clean them with sandpaper & egg brush. If hatching eggs are washed, the temperature of the water must be 43 to 44BC. The eggs should be spray washed with detergent, sanitised and fan-dried. Improperly washed eggs may rot or explode during the incubation period. Dirty eggs can be reduced if the eggs are gathered four or five times daily from nests supplied with fresh, clean nest material and if the litter in the breeding pens is kept dry so the hens feet are clean. Do not set floor eggs. Collect eggs with clean hands onto clean flats.

Select hatching eggs that are uniform in size (recommended minimum 52g for meat-type), shape and colour, with good sound shells. Do not set malformed, porous-shelled, doubled-yolked eggs or eggs with cracks. Shells that have a mottled appearance upon candling are not considered to have poor shell quality and can usually be set with good hatching results.

Since smaller eggs hatch in less time than large eggs it is recommended that eggs be separated by size and that small eggs be put in the incubator 8-18 hours after the larger eggs are set.

The hatching percentage will be the highest if eggs are held at a temperature of 16 to 17BC for not more than one week before setting. Higher temperatures initiate embryo growth. Storage temperature should be reduced to 13BC if eggs are being held for two weeks or longer. Eggs may sweat when moved to warm, humid areas. This allows bacteria to penetrate the shell. Turning eggs during the holding period is not beneficial. It has been shown that eggs held for more than 2 weeks hatched better when stored small end up (contrary to the accepted traditional large end-up position). Relative humidity should be maintained at approximately 80% in the egg holding room. Higher humidity encourages mold growth. Prior to placing eggs in the incubator, they may be removed from the egg storage room and warmed to room temperature for approximately 6 hours.

The Incubator Room

In the selection of a successful incubator room, factors such as heating, humidity, ventilation, and sanitation should all be considered. Optimum results can be expected if the temperature can be maintained at between 24 and 27BC, with uniform humidity below the level that is required in the incubator. Tropical climates (heat & humidity) make it difficult to maintain good incubator room conditions. Good ventilation and a constant supply of oxygen to remove excess carbon dioxide from the environment surrounding the incubating eggs is necessary for the developing embryo. High altitude reduces available oxygen. An incubator should not be placed near an outside wall or window in cold climates or in direct sunlight.

Incubators

Incubators are the most important equipment in the hatchery process. Many kinds of incubators are manufactured; however, the general principles of all modern machines for commercial hatchery production are the same. Incubator setting capacity ranges from approximately 14,000 to 100,000 eggs. During incubation, the hatching eggs are set vertically, with the large ends up in trays or flats in a setter and turned mechanically until about three days prior to hatching (setting period). The eggs are then transferred to a hatcher (hatching period) in a horizontal position and not turned during the hatching process. Both setters and hatchers have forced-draft air circulation, automatic temperature, humidity and cooling controls.

For small backyard poultry operators, there are small, still-air machines. Capacity of these incubators varies from 12 to 200 eggs. The eggs are set in a horizontal position and are usually turned manually. This type of incubator may be used for the entire incubation period for any kind of eggs. The source of heat is a thermostatically controlled heating element or light bulb. Humidity is supplied in most cases by water in a pan below the eggs, and ventilation controlled by small air vents or holes. Most still-air machines have transparent plastic domes through which the incubation process can be observed.

Table: *Incubation Time for some Common Avian Species*

Chicken	- 21 days
Turkey	- 28 days
Japanese quail	- 17 days
Guinea fowl	- 26 days
Pheasant, Partridge	- 24 days
Duck	- 28 days
Muscovy duck	- 35 days
Goose	- 28-32 days
Egyptian goose	- 35 days

The Principal Factors in Incubation

The five main factors affecting incubation, listed in order of importance are: temperature, humidity, ventilation, position, and turning of the eggs.

Some factors to consider before purchasing an incubator:

1. Any incubator will produce better results when operating at capacity. Egg capacity is a factor;
2. How long the incubator will last and cost of all replacement parts;
3. Service and availability of parts;
4. Accuracy of controls, to keep temperature and humidity fluctuations to a minimum;
5. The amount of labour involved in operating the machine and carrying out a thorough sanitation programme;
6. The guarantee.

Temperature

Hatching eggs may be warmed to a temperature of 25 to 30BC, prior to setting. The normal development of the embryo is dependent on the heat being held within a very narrow range in the incubator. In small still-air incubators, the temperature of the upper surface of the egg is higher than on the lower surface, while in large incubators, the air movement maintains the same temperature over the entire surface. For this reason, a still-air incubator must be operated at a higher temperature than a forced-draft incubator.

In small still-air incubators, a constant temperature of 39BC is considered satisfactory to produce good hatching results. The temperature may vary between 37.5 and 39.5BC without hurting the embryos as long as the temperature does not remain at either extreme. These readings should be taken with the bulb of the thermometre level with the upper surface of the eggs, but not in contact with the egg shell. A standing thermometre will give a more reliable reading than a hanging thermometre and the thermometre must be accurate. The temperature may rise one degree at hatching time without causing any reduction in hatch percentage.

In large incubators, the temperature, humidity, and speed of air movement are very closely dependent on each other, and since air speed varies in different incubators, it is impossible to state an exact operating temperature for all large machines, but generally it is around 37.5BC for a setter and 37BC for a hatcher. Follow the manufacturer's instructions closely with regard to temperature and ensure that instructions are for the model in use. High temperatures even for a very short period of time during any part of the incubation period will cause more harm than low temperatures.

Numerous factors may contribute to high or low incubator temperatures.

- high or low room temperatures or floor temperature under the incubator affect the operating temperature of both large and small incubators. A large incubator placed near an outside wall may have one section operating at a lower or higher temperature than the other. These problems and those listed below will result in one group of eggs hatching earlier or later than the others, increased embryo mortality or cause leg deformity in chicks & poults.
- floor temperature variation because of drains causing cold or heat under the incubator.
- an incubator thermometre not reading correctly.
- the failure of automatic equipment such as thermostats, cooling coils or automatic dampers.

- improper air circulation leading in turn to a rise in temperature in part of the incubator, because of: a decrease in fan speed, usually caused by low voltage or a slipping fan belt; incorrect spacing of filled egg trays in a partially filled incubator.
- overloading an incubator for any one setting of eggs.

Humidity

During the incubation or setting period, eggs should lose 11 to 12% of their weight (another 3 to 4% in the hatcher, after day 18), due mainly to a loss of moisture. The amount of moisture (humidity) in the incubator controls the rate of evaporation from the egg. The evaporation rate is also related to temperature, air speed, shell thickness, and size of eggs; the smaller the eggs, the greater percentage of moisture loss. Too great a moisture loss from the egg in the early stage of incubation will cause the embryo to adhere to the shell, causing death. Insufficient evaporation may cause death from lack of oxygen because of a small air cell, since just prior to pipping the shell, the embryo pips into the air cell and starts to breath air. The best guides to the correct amount of humidity in an incubator is the weight loss and the size and enlargement of the air cell during incubation, or the position at which the chick pips the shell. The degree of enlargement of the air cell should be determined by candling several eggs and estimating the averagee evaporation.

The amount of moisture in an incubator may be referred to as "relative humidity", which is a percentage of the moisture in the air at any given temperature. This can be measured by a wet-bulb thermometre.

A wet-bulb reading is based on air movement, and for this reason, a wet-bulb thermometre cannot be used to determine the amount of humidity in a still-air incubator. The relative humidity for small incubators is usually stated as 60%. Other inexpensive but less accurate equipment is available to measure relative humidity. A level water pan on the floor of the incubator under the eggs during the entire incubation period will generally supply sufficient moisture to give good hatching results, provided the moisture content of the air in the environment is neither too high or too low.

Percent egg weight loss during the setting period can act as a guide to the correct humidity. This can be done by weighing some eggs at the start of incubation and the same eggs again on the 7th and 18th day, and calculating the loss that occurred (write the weight on the shell with a pencil). Chicken eggs should lose approximately 4.5% and 11.5% during the first 7 and the first 18 days of incubation respectively.

Eggshells thicker than 0.34 mm are too thick and humidity should be reduced to increase moisture loss. Below 0.31 mm is too thin. Normal would be 0.33 to 0.34 mm at the beginning of production. Shell thickness decreases with age of breeder hen with 0.31 mm normal at 60 weeks.

In large incubators, the temperature, air speed and humidity are intimately related, and the manufacturer's instructions should be followed closely. Better hatching results may be obtained if the temperature is lowered and the humidity raised at hatching time (only if separate hatching compartments are available). Lowering the air temperature will provide an additional increase in relative humidity to keep shell membranes moist during the hatching process.

Table: *Percent Relative Humidity At Different Wet-bulb Readings*

Wet-bulb Reading (BC)	***Relative Humidity (%) at Air Temperature of***	
	37.2BC %	***37.8BC %***
32.2	70	68
31.7	67	65
31.1	65	63
30.6	62	60
30.0	59	57
29.4	56	54
28.9	53	51
28.3	51	48
27.8	48	46
27.2	45	43
26.7	43	41

Incorrect humidity may be due to a number of factors, the more important of which are listed below:

1. High environmental humidity in tropical countries.
2. A wet-bulb thermometre reading incorrectly. Remove the wick to determine whether the thermometre is reading the same as the dry-bulb thermometre.
3. Dust and dirt on the wet-bulb wick. Change wicks often and use only distilled water in the reservoir.
4. In small incubators with humidity supplied by water pan evaporation, ensure that there is always an adequate amount of water in the pan, as the rate of evaporation is dependent on the humidity in the room. The lower the humidity in the room, the higher the evaporation and vice versa. During hatching fluff settles on the surface of the water producing a film that causes

a reduction in the rate of water evaporation necessary for optimum hatchability. Clean the water pan daily and replace with clean, lukewarm water.

5. Ventilating an incubator to control temperature will lower humidity in a dry environment unless automatic humidifiers are being used in the incubator room.
6. If airflow in incubators is not uniform humidity may be different from bottom to top or side to side.

Ventilation

The free movement of oxygen, carbon dioxide and water vapour through the pores of the shell is important, since the developing embryo must be able to take in a constant supply of oxygen and release carbon dioxide and moisture. Oxygen content of 21% (present in air at sea level) and a carbon dioxide content not exceeding 0.5% in the air are considered optimum for good hatching results. Room temperature, room humidity, the number of eggs set, the period of incubation, and the air movement in the incubator all influence ventilation requirements. Ventilation problems are not the same in small incubators as they are in large incubators, where a large number of eggs are set in a very small space.

During the early part of the incubation period, ventilation in small incubators may be held to a minimum. However, during the hatching period additional ventilation must be supplied to reduce the carbon dioxide in the incubator. It is advisable not to increased ventilation until half of the hatch has been completed, since ventilating too soon will reduce the humidity. In large incubators, the manufacturer's directions should be followed, however, ventilating recommendations may not be applicable to every locality and every room condition.

If ventilation is used to control either temperature or humidity in the incubator, the control of the same factors in the incubator room are important. At a room temperature below 18 C, ventilating an incubator will reduce both temperature and humidity. In a room with high humidity, (tropical countries) the primary concern is to maintain the correct temperature.

The main ventilation consideration may be summarised as follows:

1. Ventilation is more important in large incubators than in small incubators.
2. The amount of ventilation required may be altered by atmospheric conditions.

3. Ventilation is very important in any incubator at hatching time. Insufficient ventilation may result in embryo or chick death.
4. Ventilation in excess of the recommended amount may be applied to reduce temperature or humidity.
5. The appearance of chicks panting in a hatcher at normal temperature is an indication of a rise in the carbon dioxide content of the hatcher air. Under such conditions chicks must breathe faster to obtain the required amount of oxygen and to eliminate the excess carbon dioxide. If excessive panting occurs, increase the airflow in the hatcher.

Position and Turning of Eggs

In small incubators, the eggs are maintained in a horizontal position during the entire incubation period. In large incubators eggs should be placed in a vertical position, large end up, during the hatching period. In small incubators, the eggs are moved when turned, while in large incubators they remain in a stationary position on the incubator tray and the egg tray is turned through an angle of not less than 90 in opposite directions with each turning.

The objective is the same in both types of incubators; namely, to prevent the embryo from sticking to the shell membranes. Turning also ensures a complete contact of the embryonic membranes with the food material in the egg, especially in early stages of incubation.

In small incubators, the eggs should be turned at least four times daily. It is advisable to leave some space on the tray to allow for moving the eggs forward a 1/2 turn on one turn and back a 1/2 turn on the next, thus making sure that all the eggs move.

Eggs should not be turned in a complete circle, as this has a tendency to rupture the allantois sac with resultant embryonic mortality. Wash hands carefully before turning eggs to avoid bacterial contamination of the shell. In large incubators, the trays are usually turned hourly with all the egg trays moving at one time.

For good hatchability, eggs should be turned to a position at least 45 from vertical, then reversed in the opposite direction to a similar position. In the most recent models of incubators, eggs are turned through an arc of 150 and in a few models they are turned as far as 180. The introduction of these newer methods of turning eggs has been an important aspect in improving hatchability.

Eggs should not be turned in either large or small incubators during the hatching period. The greatest benefit from turning eggs is during the first week in incubation.

Other Factors Affecting Incubation

Egg Selection

Poor quality hatching eggs do not hatch as well as eggs of good quality. The term "quality" refers to the condition outside the shell, the condition of the shell itself and that of the contents. Eggs with inferior characteristics, as discussed in "Selection and Care of Hatching Eggs," should not be set.

Sanitation

Eggs used for hatching should be clean and stored in clean containers in a sanitary egg holding room. Eggs contaminated with bacterial organisms usually do not hatch well and this poor quality is reflected in the chicks that do hatch.

Egg Handling

Rough handling of hatching eggs before they are set will increase the number of dead embryos, with mortality occurring between the 4th and 13th day of incubation. Also, jarring eggs during incubation may result in the rupture of the egg shell membrane and thereby lower hatchability.

Large fluctuations in temperature and humidity during storage will have a major adverse affect on hatchability. Refer to "Selection and Care of Hatching Eggs" for proper egg storage procedure.

Toxicity

If the interior of an incubator is painted or varnished, or if the trays are varnished, the percentage of hatch will be reduced, possibly by as much as 25%. This adverse effect disappears in about 30 days, suggesting that the ill effect is eliminated by oxidation of the paint.

This problem may be overcome without any reduction in percentage of hatch if the incubator is fumigated with formaldehyde gas at the concentration recommended for proper hatchery fumigation. The gassing should be done as soon as the paint is dry and with the incubator operating at recommended temperature and humidity for incubating eggs.

Automatic Equipment

The addition of automatic equipment has eliminated many of the problems with incubators, however, such hazards may occur when automatic devices fail. During hatching all automatic devices, such as cooling coils, automatic dampers, tray turners, alarm bells, etc., should be checked at regular intervals. Electrical failure may require that automatic equipment be reset. After an incubator is purchased, it is important to know how every part operates.

Egg Candling

Candling chicken eggs on the 7th and 18th day of incubation, may be recommended for small poultry producers. Egg candling will detect infertiles and early dead germs. Therefore, problems within the hatching flock can be identified without waiting until the incubation period is completed.

Improper Fumigation

Closely follow instructions outlined in "Fumigation Procedure". Excessive and improper fumigation can result in high mortality in developing embryos.

Fumigation of Incubators

The killing of bacterial organisms by formaldehyde gas is based on the concentrations of the gas, exposure time, temperature, and humidity of the incubator. The chemicals potassium permanganate and formalin (which is 40% formaldehyde gas) have proven to be the most effective method of destroying bacterial organisms in the hatchery. To accomplish the proper release of the gas, one and one-half parts (by volume) formalin is added to one part (by weight) of potassium permanganate. This will release the formaldehyde as gas or fumigant. When the reaction is complete, a dry, brown powder will be left. If the residue is wet, not enough permanganate was used; if the residue is purple, too much permanganate was added.

Concentration

The recommended concentration for effective fumigation is 53 mL of formalin added to 36 g of potassium permanaganate per cubic metre of space to be fumigated, or 1 1/2 mL of formalin added to 1 g of potassium permanganate per cubic foot of space to be fumigated.

Caution: Never add the permanagante to the formalin. Heat is generated when the two chemicals are combined, and care should be taken. Formaldehyde gas is generated quickly. Do not allow the fumes to get into the eyes. Personnel should use a respirator or wear a mask to avoid unnecessary exposure. Ventilate the incubator room to remove fumes that escape from the incubator.

Time

It is not recommended to fumigate setters with hatching eggs in them, but if such treatment becomes necessary, embryos between 24 and 96 hours of age should not be exposed to the above concentration of formaldehyde. Hatching compartments should be fumigated after the eggs are transferred from the setter to the hatcher, again after the hatch has been taken off and before the refuse has been removed from the trays, and finally after the hatcher has been thoroughly cleaned. Do not fumigate chicks with this concentration of formaldehyde gas. Small still-air incubators should be fumigated after the chicks have been removed and prior to discarding the refuse from the tray and again after the incubator has been thoroughly cleaned.

Fumigation Procedure

1. Make sure the temperature and humidity of the incubators are at normal operating conditions.
2. Measure the inside volume of the machine in cubic feet or cubic metres (length x width x height).
3. Close the ventilators, but leave the fans on.
4. Weigh the required amount of potassium permanganate into a wide enamelware or earthenware vessel large enough to accommodate the boiling and splattering action experienced when the formalin is added. Place the vessel and the permanganate in the area to be fumigated; then add the formalin.
5. Close the door immediately and leave closed for 20 minutes.
6. After 20 minutes, open the ventilators.
7. Open the doors of the machine for five minutes, leaving the fan on to allow more of the formaldehyde gas to escape, or neutralise it with a 25% solution of ammonium hydroxide equal to one-half the amount of formalin used. The hydroxide should be thrown directly on the floor of the machine and the doors closed. The formaldehyde gas will quickly be neutralised.

Continuous Fumigation of Hatcher

The greatest increase in bacterial organisms occurs during the hatching period. These can be reduced, but not completely eliminated, by slow release of formalin in the hatcher during the last 48 hours of the hatching period.

For continuous fumigation to be effective, hatching eggs should also be fumigated at transfer time with the recommended concentration.

Place formalin in pan about 25 mm in depth, allowing 58 cm^2 of pan for each cubic metre of hatcher space, or a pan about 1 in. deep, allowing 30 in2 of pan for approximately 1000 ft3 of hatcher space. Do not use permanganate. The pan should be placed in the open area of the hatcher in direct line with the airflow.

Place the pan of formalin in hatcher 48 hours prior to hatch completion. To overcome fluff deposited on the formalin, add more formalin about 24 hours before hatch is complete to increase evaporation.

If the fan stops (mechanical failure, etc.) and reduces the airflow, remove the pan of formalin immediately. If hatchers are only partially full, continuous fumigation is not recommended, as a reduction in airflow will create an excessive build up of formalin in the hatcher.

Effects of Fumigation

1. Properly carried out, fumigation should not affect hatchability.
2. Fumigation will only kill bacteria that are present on the surface of hatchery refuse. Fumigation will not kill bacteria inside unhatched or pipped eggs. It is important to dispose of hatchery refuse carefully to minimise hatchery contamination.
3. The hatchery room must be separate from the tray dumping room and from the chick processing area. Air flow and traffic must be controlled to prevent contamination of the chick processing and holding areas.
4. Efficient fumigation along with other sanitary measures should control navel infection (omphalitis).
5. Fumigation is not intended to replace a thorough cleaning programme.

Hatchery Design

Hatchery layout is important to a good sanitation programme. Arrange the hatchery so there is a one-way flow of material from the point where the hatching eggs are brought in to where the processed chicks go out.

Hatcheries should be designed so the flow of eggs, chicks and personnel does not spread contamination from one room to the next. Doors, including one-way doors, help stop cross contamination between rooms. Positive air pressure prevents contamination through an open door. Workers should change outer clothing and wash hands before moving from one work station to another.

Summary of Some Important Factors

1. Feed breeder flock hatching ration that is well fortified with essential nutrients.

2. Use healthy breeding stock.
3. Provide good egg-holding facilities.
4. Avoid holding eggs in storage for more than one week.
5. Prewarm eggs for 6 to 8 hours at incubator room temperature.
6. Set clean, good quality eggs. Delay setting small eggs (those more than 10% less than average) for 8 to 16 hours.
7. Maintain correct incubation temperature, humidity and ventilation. Make sure air intake does not draw contaminated air into the incubator.
8. Turn hatching eggs frequently.
9. Maintain incubator room temperature between 21 and 24BC with good ventilation and relatively high humidity.
10. Fumigate regularly.
11. Clean vaccinating and beak trimming equipment. Newly hatched chicks may pick up contamination and infection in the hatchery from vaccinating and beak trimming equipment. This equipment requires a very rigid sanitation schedule.
12. Practice strict sanitation; cleanliness is very important for successful hatching operation. Make sure belts, equipment and workers hands used to move eggs or newly hatched chicks are kept clean.

Animal Welfare Problems in Chickens and Other Poultry Caused by Single Trait Genetic Selection for Meat and Egg Production

Chickens have several serious welfare problems that come from bad genetics and can be fixed only with good genetics The biggest problem in many intensively raised animals is pushing the animal's biology for more and more production. Breeders choose the most productive animals — the fastest growing, the heaviest, the best egg layers, and so on — and selectively breed just those animals. Bad things always happen when an animal is overselected for any single trait. Nature will give you a nasty surprise.

Bone breakage is a very serious problem in both caged and cage-free hens because laying hens have been overselected for egg production. Commercially bred hens put all their calcium and minerals into forming eggshells, and their own bones become depleted. Their bones are so weak that in cage-free systems a hen can break her leg just jumping off her perch. The only way to solve this problem is for the industry to accept the fact that birds with strong bones will produce slightly fewer eggs.

Laying hens have other problems, too, especially feather pecking and cannibalism. Feather pecking is what it sounds like: one hen pecks at another hen's feathers or pulls a feather out all the way. Severe feather pecking can lead to cannibalism, with the victim hen being wounded and then killed by the hen doing the pecking. Even though a feather-pecking hen can kill her victim, feather pecking probably isn't driven by the RAGE system. We know this because of studies mixing unfamiliar hens together. Aggression goes up, but feather pecking and cannibalism don't. They aren't the same thing.

Instead, feather pecking is probably displaced or redirected SEEKING behaviour. It's a kind of foraging or exploration of another bird instead of the ground. We know this because of research showing that chickens housed on litter do a lot less of it. They peck at the litter on the ground, not at each other's feathers. The more active the bird and the more foraging behaviour it does naturally, the more likely it is to develop severe feather pecking. Both feather pecking and cannibalism are affected by genetics.

Some modern broiler chickens have genetic problems related to growth. I was shocked to learn at a chicken-breeding seminar that the broiler chicken has been so overselected for rapid growth that its bone physiology is totally abnormal. In normal bone development, the body first "erects" a scaffolding or frame of cartilage and then fills in the frame with minerals that harden into bone. After the bone has hardened, the cartilage dies off through programmed cell death. In broiler chickens, something goes wrong with the cartilage, so the bones don't have support while they're hardening and end up misshapen. I liken it to building a new basement wall and taking down the plywood concrete forms before the concrete has fully hardened. In some of the worst cases, a chicken's feet are rotated almost 90 degrees and the legs are twisted. These chickens are genetically lame. Several studies have shown that lame broilers will choose feed laced with painkiller over their regular feed, and a study of lame turkeys showed that they started moving around a lot more once they were on painkillers. The industry has created chickens that have chronic pain in order to get birds that grow at the far outer limits of what is biologically possible. When an animal's biological system is pushed to the point where the physiology is totally pathological, I get disgusted.

The other problem is that modern broiler chickens have been bred to have stupendous appetites so they'll grow super-fast and reach market weight as soon as possible. The trouble is that the breeder chickens, the parents of the broilers, have the same stupendous appetites as their chicks. If you let a broiler breeder chicken eat everything she wants, she will become obese, her fertility will decline, and her life will be shortened. These chickens have to be kept on a strict diet just to maintain normal weight. They act miserable, and many of them develop stereotypies. These birds have low welfare no matter what you do. If you let them eat all they want, they have bad welfare and if you don't let them eat all they want, they also have bad welfare. It's terrible. The industry is going to have to breed parent stock with smaller appetites. There's no other way to fix the problem.

Then there are other genetic problems that no one understands. One of the worst cases was the rapist roosters. I wrote about them in Animals in Translation. Fortunately, the broiler industry has made some genetic changes to correct these problems, although there's still a way to go. The rapist roosters violently attack hens and injure and even kill them. Before the 1990s there weren't any rapist roosters. They just suddenly appeared out of the blue. First it was just one strain of roosters that had become aggressive but within a couple of years almost all strains had developed the same behaviour. Nobody knows why.

The rapist roosters have two problems: They are hyperaggressive and they have stopped doing the courtship dance the hen needs to see before she will mate. They've lost the little piece of genetic code that makes them do the dance. When the hens don't see the courtship dance, they don't become sexually, which may make the roosters' aggression worse. An unreceptive hen would be a form of frustration because it is a restraint on the rooster's action. So the RAGE system would be activated to some degree.

When I wrote Animals in Translation it looked like the rapist roosters were a side effect of the industry's selective breeding programme to create chickens with bigger breasts for more white meat. But now researchers aren't sure what caused it, or whether the hyper-aggression and the bad courtship behaviour are the same problem or two different problems that happened at the same time. Industry breeding programmes are trade secrets. It's obvious the industry is selectively breeding for larger breast size because breast size is getting larger. But we don't know what other selective breeding programmes the industry might be using.

Ian Duncan has an interesting theory about what might have happened. Dr. Duncan points out that big-breasted male birds have trouble mating because their huge chests get in the way. Male turkeys have such big breasts now that they can't mate at all and the hens have to be artificially inseminated.

Dr. Duncan says that if the same thing is happening to male chickens, the broiler breeder industry may have misdiagnosed the problem. When broiler breeders see chickens with decreased fertility, they attribute the problem to low sex drive. It's possible that the breeders who created rapist roosters were actually trying to increase roosters' sex drive. If the breeders selected for higher libido they could have mistaken a little bit of aggressive behaviour towards the hen for higher sex drive and ended up breeding hyper-aggressive roosters that for some reason had also lost their courtship dance. We'll probably never know.

Today the aggressive rooster problem has been greatly reduced although it hasn't been eradicated.

Better Breeding Strategies Using Group Genetic Selection

Most of the time breeders deal with genetic problems by culling chickens that have the problems and mating the ones that don't. Another interesting approach is group selection. The researcher Bill Muir at Purdue University has shown that you can reduce feather pecking genetically by using a technique called group selection. With group selection, instead of picking certain individuals, you pick certain family groups to breed. Dr. Muir has done this by raising several "sire family" groups — groups of chickens related to each other through their father — and then selectively breeding the group that has the highest egg productivity and the lowest amount of feather pecking and cannibalism.

Group selection has a couple of advantages over individual selection. First, the fact that you're working with groups instead of individuals means you know something about the birds' behaviour in a group. When breeders select high-productivity, individually housed laying chickens to breed, they don't know whether they are feather peckers or not because they've never lived with other birds. Individually housed laying hens can't express the behaviour.

Second, when you have a group of genetically related chickens living together you also see how living in a group affects their behaviour and productivity. When you choose which group to breed, you're not just choosing one genetic strain of chickens over another, or two desirable behavioural traits (good egg laying and low feather pecking) over undesirable traits. You're choosing one way of relating to the environment over another.

This is important because most behaviour is affected by what's going on in the environment. A hard-wired behaviour like the courtship dance is always the same no matter what's going on in the environment. It's like a computer subroutine; once you turn it on it just runs. But everything else is affected by the environment. When breeders use group selection instead of individual selection, they're factoring in the way the selected group relates to its social and physical environment.

Today, only a handful of companies provide all of the commercial layers and broilers around the world, which has greatly narrowed the gene pool. This has created a risky situation because genetically similar animals are vulnerable to the same diseases. Sure enough, when the Australians phased out their homegrown broilers and imported American birds, they ended up with more disease problems.

This is why it's important to preserve the old breeds of animals and poultry. Keeping the classic breeds alive is the only way to preserve genetic diversity and to save animals that have valuable genetic traits breeders may want to breed back into commercial lines in the future. The meat from some of the old breeds is more tender and better quality than meat from animals bred for rapid growth, and the chickens are hardier, too. They perform better in pasture-based or organic farms. They are beautiful, unique animals that shouldn't be destroyed by commercial breeding. Fortunately, many of the older breeds of poultry and livestock are being raised by local farmers and sold in farmer's markets or to gourmet restaurants. If a serious disease ever kills commercial broilers or layers, the entire world will be thanking the small producers and hobbyists who have kept the old breeds of chickens from becoming extinct.

Chapter 2

Pre-Breeder Diets

While most breeding companies provide nutrient specifications for pre-breeder diets, there is considerable variation in their commercial use and application. Using a pre-breeder or pre-lay diet is based on the assumption that the birds nutrient requirements change in this critical period of the birds life. There are certainly major changes occurring in the birds metabolism, hopefully related to ovary and oviduct development, and so this is the basis for a specialised diet at this time. With egg-laying stock, pre-lay diets most often involve a change in calcium nutrition, in order to establish the birds calcium reserves necessary for rapid and sudden onset of eggshell production. The same situation can be applied to heavy breeders today, because with flocks of uniform body weight and with good light management, the subsequent synchronisation of maturity leads to rapid increase in egg numbers up to peak production. However, most often pre-breeder diets are used in an attempt to "condition" or correct growth and/or body compositional problems that have arisen during the 14-18 week growing period. In these situations managers are perhaps ill-informed of the expectations of merely changing diet specifications at this time.

The Pre-Breeder Period

Although there is no specific pre-breeder "period", most consider the 19-23 week period to be the major transition time for sexual development of the bird. During this time (4 weeks) the pullet is expected to increase in weight by about 570 g. This is somewhat more than the growth expectation of around 340 g for the previous 4 weeks (15-19 week) or growth of around 470 g for the 4 weeks from 23-27 weeks of age. It is expected that a significant proportion of this growth spurt will be as ovary and oviduct, which are developing in response to light stimulation. A practical complication of this expected development, is that it invariably coincides with move of the pullets from grower to breeder facilities. Under adverse conditions, such as transportation over long distances, heat stress, etc., then birds can lose up to 100 g of body weight at this critical time. If weight loss is characteristic of such transportation, then pullets should be given an extra feeding. For example, pullets should be moved on an "off-feed" day, but be fed that day in the breeder house after all birds are housed. We cannot allow weight loss at this critical time, and so the question to be answered is - do pre-breeder diets help in this physical move, as well as prime the bird for sexual maturity? Development of the ovary and oviduct require both protein/amino acids and energy (fat) accretion. Nutrients of interest, therefore, are protein and energy, together with increase in calcium for early deposition of medullary bone. However it has never been clearly established that such nutrients need to come from a specially fortified diet versus simply increasing the feed allowance of the grower diet or breeder diet that is introduced prior to maturity.

Calcium Metabolism

With egg-layers, pre-lay diets are used essentially to pre-condition the pullet for impending eggshell production. The very first egg represents a 1.5-2.0g loss in calcium from the body, the source of which is both feed and medullary bone reserve. Today breeder hens are capable of a sustained long clutch length which is necessary to achieve the 85-87% peak production that is now readily attainable. Calcium metabolism is, therefore, very important for the breeder. With Leghorn hens the consequence of inadequate early calcium balance is cage layer fatigue. Breeders do not show such signs, because they naturally have more exercise, and also have a readily available rich source of diet calcium in the form of their flockmates eggs. Hens have an innate ability to select out calcium, and so improperly fed breeders will eat litter and eggshells in an attempt to balance their diet. However inadequate calcium in the diet does lead to disruption of ovulation, and so these birds stop laying until their meagre calcium reserves are replenished. In a breeder flock, it is the larger bodied, early maturing pullets that are disadvantaged in this manner.

Commercially, we see three different approaches used in pre-breeder calcium nutrition. Firstly, is the use of grower diets that contain just 0.9 - 1.0% calcium being fed up to 5% egg production. This is the system that was used many years ago, and unfortunately is still sometimes used today. At 5% egg production, we do not have 100% of the flock producing at 5% egg production - rather we have closer to 5% of the early maturing pullets producing at close to 100% production. Pullets can produce just 2-3 eggs with a diet containing 1% calcium. After this time they will eat litter/eggs as previously described, or more commonly simply shut down the ovary. With this approach, birds may in fact be at 10-15% production before the breeder diet is introduced, because no farm system allows for instantaneous change in feed supply as feed tanks are hopefully never completely empty. There is no justification, therefore, for this old system of feed management, because it will be very detrimental to life-time productivity of today's genetic stocks.

The second system involves the classical pre-breeder diet containing around 2% calcium, which is really a compromise situation. It allows for greater medullary bone reserves to develop, without having to resort to the 3.5% calcium as used in a breeder diet. However 2% calcium is still inadequate for sustained eggshell production - with this diet the breeder can produce 4-6 eggs before ovulation pattern is affected. If a pre-breeder diet is used, therefore, and a moderate calcium level is part of this programme, then the diet must be replaced by the breeder diet before egg production starts. A good rule of thumb is to change from pre-breeder to breeder when the very first egg is noticed, because this occurs usually around 10 days before 1% egg production.

The third option is perhaps the most simple solution, and involves changing from grower to breeder at 1st egg (10 days before 1% production). Having the breeder diet in place before maturity, ensures that even the earliest maturing birds have adequate calcium for sustained early egg production. Proponents of pre-breeder diets suggest that breeder diets introduced early provide too much calcium, and that this contributes to kidney disorders, because the extra ingested calcium must be excreted in the urine. There is an indication with Leghorns that feeding adult layer diets for 10-12 weeks prior to maturity can adversely affect kidney function, especially if birds are also challenged with infectious bronchitis. However feeding "extra" calcium for one or two weeks prior to maturity has no such effect. It is also interesting to realise that most roosters today are fed high-calcium breeder diets, which provide 4-6x their calcium needs, yet kidney dysfunction is quite rare in these birds.

Body Weight and Size

Body weight and body condition of the bird around the time of maturity, are perhaps the most important criteria that will ultimately influence breeder performance. Body weight and body condition should not really be considered in isolation, although at this time we do not have a good method of readily assessing body condition. Each strain of bird has a characteristic mature body weight that must be reached or surpassed for adequate egg production and egg mass output. In general, prelay diets should not be used in an attempt to manipulate mature body size. The reason for this is that with most flocks it is too late at this stage of rearing to meaningfully influence body weight - all too often prelay diets are used as a crutch for poor rearing management.

However, if birds are underweight when placed in the breeder house, then there is perhaps a need to manipulate body weight prior to maturity. Under controlled environment conditions, this can sometimes be achieved by delaying photostimulation. If prelay diets are then necessarily used in an attempt to correct rearing mismanagement it seems as though the bird is most responsive to energy. This fact likely fits in with the effect of estrogen on fat metabolism, and the significance of fat for liver and ovary development at this time. While such high nutrient density pre-lay diets may be useful in manipulating body weight, it must be remembered that this late growth spurt (if it occurs) will not be accompanied by any meaningful change in skeletal growth. This means that in extreme cases, where birds are very small in weight and stature at say, 16-18 weeks of age, the end result of using high-nutrient dense prelay diets may well be pullets of correct body weight, but of small stature. These short-shank length pullets seem more prone to prolapse/pick-out, and so this is another example of the limitations in use of classical pre-lay diets.

Use of high-nutrient dense prelay diets to manipulate late growth of broiler breeder pullets does, however, seem somewhat redundant. The reason for this is that with restricted feeding programmes, it is more logical to increase feed allowance than to add the complexity of introducing another diet. The only potential problem of this programme is that in extreme cases feed intake is increased to a level that is in excess of the initial allowance of breeder diet at start of lay ie. ensure that breeders are not subjected to a step-down in feed allocation at time of first egg.

Body Composition

While body composition at maturity may well be as important as body weight at this age, it is obviously a parameter that is difficult to measure. There is little doubt that energy is likely the limiting nutrient for egg production for all classes of birds, and that around peak production, feed may not be the sole source of such energy. Labile fat reserves at this time are, therefore, essential to augment feed sources.

These labile fat reserves become critical during situations of heat stress or general hot-weather conditions. Once the bird starts to produce eggs, then its ability to deposit fat reserves is greatly limited. Obviously if labile fat reserves are to be of significance, then they must be deposited prior to maturity.

Egg Weight and Hatchability

It seems as though egg size is ultimately controlled by the size of the yolk that enters the oviduct. In large part this is influenced by body weight of the bird, and so factors described previously for body weight can also be applied to concerns with egg size. There is a general need for as large an early egg size as is possible. Most attempts at manipulating early egg size have met with limited success. Increased levels of linoleic acid in prelay diets may be of some use, although levels in excess of the regular 1% found in most diets produce only marginal effects on early egg size. From a nutritional standpoint, egg size can best be manipulated with diet protein, and especially methionine concentration. It is logical, therefore, to consider increasing the methionine levels in prelay diets.

For breeders we must also consider egg composition as it relates to early hatchability success. Eggs from young breeders seem to inherently have a hatchability problem, and perhaps this is one of the reasons that we wait for egg size to increase before sending eggs to the hatchery. The reason for this early hatch problem is not fully resolved, but most likely relates in some way to maturity and development of embryonic membranes and their effect on transfer of nutrients from the yolk to the embryo. However part of this problem may also elate to inadequate transfer of vitamins into the egg. For a number of critical B-vitamins, their concentration in successive eggs does not plateau until after 7-10 eggs have been laid. The effect of pre-lay nutrition on these factors probably warrants further study, but at this time these problems cannot be resolved by simply over-fortifying pre-breeder diets with vitamins or certain fatty acids.

Conclusions

Pre-breeder diets can successfully be used as part of a feeding programme aimed at maximizing production potential in young breeders. However any desired increase in nutrient intake prior to maturity can most easily be achieved by simply increasing the feed allowance of either grower or adult breeder diet at this time. If pre-breeder diets are used, then 19-23 weeks seems the most ideal time, assuming 1% production will occur around 24 weeks of age.

Time to Feed Your Breeders

What is the optimum time of day to feed growing and adult hens and roosters? As with most questions, the answer is "it depends" and this situation certainly applies to feeding breeders. For growing birds the variables are ability to observe feeding behaviour, and potential effects of heat stress. With adult birds we have the added factors of eggshell quality and conflict of time associated with mating or nesting.

Growing Birds

For growing birds, feed is eaten in a very short period of time (30 minutes - 2 hrs. depending upon age and frequency of feeding) and so choice of feeding time has little real effect on other daily activities. In fact feeding and drinking are the major activities of the immature bird. Most producers will feed growing pullets and roosters early in the morning, especially in warm or hot climates. Digested feed is not utilised with 100% efficiency, and a by-product of such inefficiency is heat production in the birds body. In most situations this extra heat (sometimes called heat of metabolism, specific dynamic action, or heat increment), peaks about 4-6 hours after feed is eaten.

Because of the restricted feeding programme, feeding time is short and predictable, and so the heat of metabolism will consistently peak 4-6 hrs. after feeding time. In hot climates peak environmental heat load occurs in the early afternoon, and so there is a distinct disadvantage to having extra heat generated in the birds body at this time. For this reason we have the common practice of feeding birds at 6-7 am. With such early morning feeding, the heat load of nutrient metabolism occurs before the early afternoon daily high temperature. Alternatively, growing birds could be fed in late afternoon or early evening. However this latter situation does not work well with short-daylengths for growing birds.

With mechanical feeders there is a tendency to feed birds even earlier, sometimes at daylight or when artificial lights are switched on. There are two disadvantages to very early morning feeding. Firstly feeding often occurs before staff are present to observe feeding activity and bird distribution. Under these conditions it is impossible to know if feed is being evenly distributed and if all birds have access to the feed.

The second problem, which becomes more acute as birds get older, is the condition of choking, which occurs with a small percentage of older birds, especially on every-day feeding. This problem can often be resolved by switching on drinkers for at least on hour before feed is available. This is obviously impossible to accomplish if birds are mechanically fed at first daylight or when artificial lights are switched on - pullets seldom drink in the dark period.

The ideal feeding time for growing pullets and roosters therefore is early morning, when staff can observe feeding behaviour, and after birds have had access to water for up to 1 hour.

Adult Breeders

Choice of feeding time of adult breeders can influence the production of settable eggs, eggshell quality, fertility and hatch of fertiles. In most instances these factors are a consequence of feeding activity displacing other important daily routines, such as nesting and mating. Breeder hens consume their feed in 2-6 hours each day. This large variation in feed clean-up time relates to diet energy level, feed texture and perhaps most importantly, environmental temperature. In hot climates breeders often take much longer to eat feed, and this is especially true of high-yield strains. Most managers consider this extended feeding time to be advantageous, because it ensures more even allocation of feed across the flock where even the most timid birds have time to eat.

If breeders are fed early in the morning, then most intense feeding activity will be over by 9 a.m. Again this is ideal in terms of reducing heat load in the early afternoon period. This timing is also ideal in terms of differentiating the main feeding time from nesting activity. Depending upon when lights are switched on in the morning, most eggs are laid in the 9 am - 12 noon period. Feeding at, say 8 am, would, therefore, induce birds to feed at a time when they are usually in the nests. In fact eggs dropped in the area of the feeder are a very good indication of late-morning feeding. Obviously some of these eggs will get broken or become too dirty for setting.

A few years ago there was interest in feeding breeders in the late afternoon. The main advantage is claimed to be an improvement in eggshell thickness, and in fact in many field trials this is found to be true. Improved shell thickness is likely a consequence of the bird eating calcium at a time when shell calcification is starting (for the next days egg) and also the bird having more feed (with calcium) in its crop when lights are switched off. If eggshell quality (thickness) is a problem, then afternoon feeding seems a viable option. Alternatively, birds could be given a "scratch" feed of large particle limestone or oystershell in the late afternoon.

However, late afternoon feeding has a number of potential disadvantages. Firstly there is increase in shell thickness. This should not be a problem as long as incubation setter conditions are adjusted so as to maintain normal moisture loss. In most situations this means reduction in setter humidity to account for less moisture loss through a thicker shell.

A greater concern with later afternoon feeding is potential loss of mating activity, and increase in incidence of body-checked eggs. Mating activity is usually greatest in late afternoon. If hens are more interested in feeding at this time, then there can be reduced mating activity and also more aggression between males. Body-checked eggs are characterised by a distinct band of thickened shell around the middle of the egg (sometimes called belted eggs). This defect is caused by the eggshell breaking during its early manufacture in the birds uterus. The bird repairs the crack, but does so imperfectly. Such eggs have reduced air and moisture transfer characteristics, and usually fail to hatch. The most common cause of body-checked eggs is sudden activity, movement, stress etc. on the bird. This extra activity takes place when feed is given in late afternoon, and so there will likely be fewer settable eggs produced.

Conclusions

Early morning feeding in breeders is usually recommended because all associated factors and consequences of this practice are positive for the bird and the production of settable eggs. The only concern is with mechanical feeders where there is a temptation to feed too early in the morning, and before staff are present to observe bird activity.

Challenge Feeding

High and sustained peak egg production can only be achieved with uniform breeder flocks fed to meet their nutritional requirements. With 85-88% peaks now possible in the industry, it is obvious that we have to carefully plan and execute a feeding programme tailored to meet the breeders' nutrient needs. Underfeeding results in very short duration peaks, of only 3-4 weeks, and those are usually associated with the classical sign of loss or stall-out in body weight for 1-2 weeks. On the other hand overfeeding, especially with energy, will result in excessive weight gain, and while peak production may be little affected, there will be precipitous loss in egg production through 34-64 weeks of age. The basis of feed allocation at this critical time is obviously to allow genetic potential for increases in both egg numbers and egg size, and also to allow for modest weekly gains in body weight. Managers should consider "challenge feeding" as part of their feed management system at this critical time.

Challenge feeding involves giving the hens extra feed on 2 or 3 days each week, based on need, without changing the base feed quantity scheduled for the flock. For example, a flock may receive 168 g/bird/day at peak, with an additional "challenge" of 7 g/bird/day given three days each week. The challenge feed is, therefore, equivalent to 3 x 7g / 7d = 3 g/bird/day. In reality birds receive the equivalent of 168 g + 3g = 171 g/bird/day.

The immediate question is why bother with this more complicated system, and just give the flock a base feed allowance of 171 g/bird/day?

The advantages of challenge feeding, rather than simply increasing the base allocation are:

1. On days of challenge feeding, feeding time will increase, and this helps to improve uniformity.
2. It is easier to make adjustments to nutrient intake based on day-to-day change in needs as may occur with changes in environmental temperature.
3. Birds become accustomed to change in feed allocation, which will be important once feed withdrawal is practised after peak.
4. Ease of tailoring nutrient needs to individual flocks. For example, a base feed allocation of 165-175 g/bird/day may be standardised across all flocks, with individual flock needs at peak being tailored with the quantity and/or frequency of challenge, depending upon actual production, environmental temperature, etc.

The actual quantity and timing of challenge feeds must be flexible if they are to be used efficiently. In practice the challenge should not represent more than 5% of total feed intake, and most often the quantity will be 2-4%. On the other hand, the quantity of the challenge should be large enough to meaningfully contribute to the factors listed previously. For this reason there needs to be a balance between the quantity of feed given, and the frequency of this feeding. For example, a daily challenge of 3 g/bird/day given daily will be much less effective than 7 g/bird/day given 3 times each week. In both instances birds are receiving 21 g/week as a challenge, but in the later example the challenge quantity is more meaningful and we are more likely to see a bird response in terms of egg output.

Challenge feeding should start when birds are at 60-70% production, and should be discontinued when egg production falls below 80%. For most flocks, therefore, we can expect to practice challenge feeding from about 29 through 40 weeks of age. The idea of challenge feeding is to more closely tailor feed allocation to breeder hen needs, and so there should be no standardised system. Managers must be given flexibility to alter challenge feeding based on fluctuating needs. In most instances the challenge will be used to lead bids into a sustained peak. Because the concept of challenge feeding is to more closely tailor feed allocation to needs, then it is usual practice to alter the quantity and/ or duration of challenge as birds progress through peak egg production. Maximum challenge feeding should coincide with peak egg output, with lesser quantities given prior to, and after actual peak. On this basis we recommend challenge feeding to be reduced (but not discontinued) once birds are 2% below peak egg production. Following are three examples of challenge feeding tailored to three different flock situations.

Challenge feeding can also be used post-peak if there are precipitous declines in egg production related to minor disease challenge or management or environmental stress. Under these conditions, challenges of 10 g/bird/day for two consecutive days are recommended. If no immediate response is seen in egg production, then challenge should be discontinued. If egg production returns to normal, then the challenge should gradually be reduced over the next 2-3 days.

Challenge feeding allows tailoring of feed allocation to suit individual flock needs. Managers should be flexible in actual allocations, although maximum challenge feed allocation needs to coincide with peak egg production. Breeders will respond to a good challenge programme, with sustained peak production and better post-peak persistency. On the other hand, the challenge should not usually represent more than 5% of total feed intake, because excessive challenge will invariably result in obesity and related loss in post-peak performance.

In general, when birds are subjected to such stresses as variable feed quality, mycotoxin challenge and/or fluctuating or extreme environmental temperature, then a high base feed allowance, coupled with aggressive feed challenge, is recommended. On the other hand, lower feed inputs are possible where consistent quality high-energy feeds are used, and where there is good environmental control.

Spray Sanitizing Hatching Eggs

Shell surface contamination of the hatching egg is inevitable in the breeder house environment. Normally, hatching eggs are not immediately set in the incubators upon lay but are one to ten days old before being set. During hatching egg storage there will be multiplication of any bacteria already on the shell surface. The greater the number of bacteria on the eggs, the greater the chance of those bacteria invading the interior of the egg. The goal of the breeder house manager is to use good management strategies to minimise shell contamination. These strategies include such practices as:

- Keeping nests clean.
- Keeping storage facilities clean.
- Collecting eggs frequently.
- Maintaining proper temperature and humidity in the egg room.

In addition to good management practices, procedures for sanitizing shell surfaces can prevent microbial invasion if used properly. One method used extensively in the past was fumigating hatching eggs with formaldehyde gas after collection.

This procedure proved very effective but caused the Occupational Safety and Health Agency (OSHA) to express concerns for safety. An alternative to fumigation is spraying the shell surface with an egg shell sanitizer. Numerous sanitizers on the market have proven useful, but, to be effective, the sanitizer must be used correctly.

Proper Selection and Dilution

The use of an effective sanitizer is essential to minimise contamination of the shell surface. Different classes of sanitizers (Quaternary Ammonium, Phenolics, Peroxides, etc.) kill microbes in different ways. Therefore, to assure effective control, the specific procedures for each sanitizer must be followed meticulously. Some classes of sanitizers are more effective than others when used in adverse environmental conditions like poor water quality. A sanitizer used on the egg must be effective at controlling microbial populations yet not toxic to the developing embryo.

The entire sanitizer formulation should be examined to assure that none of the other compounds in the formulation can have an negative effect. Even compounds which are normally used as a hatching egg spray can be toxic if they are used at a concentration greater than the manufacturer recommendation. Sanitizers have been evaluated for optimal dilution, and using a concentration above the recommendation could potentially harm the embryo. Care must also be exercised not to use a compound which can deter the movement of oxygen to the embryo through the shell.

Implications of Improper Sanitizing

Proper selection and use of a sanitizer is essential to good spray sanitation management and can prevent additional problems in the hatchery. Since most incubators have greater than a 40,000 egg capacity, thousands of eggs and chicks could become contaminated if an infected egg explodes, breaks or becomes cracked inside the incubator.

The proper practice of good management strategies will prevent microbial outbreaks and aid in the production of quality chicks.

Contamination of Hatching Eggs

Where does Microbial Contamination Come from?

Bacteria and mold which can affect hatching eggs are found everywhere in the environment - in soil, in manure, and even on the dust particles in the air. The most common way hatching eggs become contaminated is by allowing fresh eggs to lay in dirty nests or on the floor and slats.

How Does Microbial Contamination Affect the Eggs and Chicks?

When a large number of bacteria get onto the shell surface, the chances of bacteria invading the egg increase. Bacteria inside the egg may use the nutrients found in the egg to multiply, robbing the embryo of a crucial food source or perhaps producing a toxin harmful to the embryo. During incubation, bacteria can actually prevent embryonic development, ultimately causing the embryo to die. Even if the embryo of a contaminated egg survives hatching, the chick will either die in the broiler house or simply not grow as it should.

Contaminated eggs that fail to hatch in the incubator can also affect other, healthy eggs. If one contaminated egg should crack in the incubator, it may spread bacteria to other eggs or newly hatched chicks. In fact, one egg can affect an entire incubator.

Does the Egg Possess any Anti-microbial Defences?

Although bacteria and mold can easily find a path into a cracked egg, the intact egg possesses many defences that prevent microbial infection. Barriers that protect the egg include the cuticle, the shell, the shell membrane and the albumen or egg white.

A protein layer located on the shell surface is called the "cuticle." The cuticle helps occlude, or cover over, some of the pore openings to minimise bacterial penetration. Sometimes, though, if the cuticle layer is thin, the pores are too large or the shell too thin, bacteria can enter through the pore of the shell.

If this should happen, the shell has two membranes lining the inside of it which act like a filter to prevent penetration. Additionally, the albumen contains natural compounds which can kill any bacteria that may evade the egg's other protective features. However protected the egg may seem, if the number of bacteria is too great, the natural defences cannot prevent invasion. Good management practices are essential to minimise the assault of bacteria upon freshly laid hatching eggs.

What Management Strategies Can Be Practiced to Reduce Egg Contamination?

- Collect eggs frequently to minimise the time that they are exposed to a contaminated environment.
- Keep egg laying areas as clean as possible, including the nest litter or pads.
- Remove eggs to the egg cooler as soon as possible after lay; cooler temperatures will slow the growth of bacteria on the shell surface.
- Prevent moisture from accumulating on the shell. Moisture provides a needed nutrient for microbial growth and might also provide a medium to aid the movement of microbes through the shell.
- Use authorised egg shell sanitation or fumigation programmes properly.
- Minimise the number of cracked or broken eggs. Egg contents can provide nutrients for the bacteria to multiply and spread.
- Avoid abrasive cleaning of the egg which can affect the integrity of the shell.
- Increase efforts to minimise contamination as the breeder flock's age increases. The shell becomes thinner with age and more prone to bacterial infection.

Profiling Egg Storage: The Effects on Egg Weight Loss and Hatchability

Background

Cool storage of hatching eggs is a "necessary evil" which hatchery managers have to endure. When coordinating the incubation of millions of eggs it is sometimes necessary to store hatching eggs, there-bye putting the development of the chick embryo "on hold". This storage does not come without a cost. Hatchery managers are aware of the general "rule" that the hatchability of eggs stored more than one week is reduced. There is no data on modern strains to show how much hatchability declines as storage progresses. The present study was conducted to fill this information void.

Approach

Broiler breeder eggs (4032) were collected from a commercial flock at peak production. The eggs were divided into 8 groups of 504 eggs and stored for 2, 4, 6, 8, 10, 12, 14, or 16 days. A portion of eggs from each storage group was weighed regularly during storage. All eggs were incubated and the fertility, hatchability and incidence of embryo death calculated.

Results and Discussion

Fresh egg weight was not different between any of the 8 egg storage durations. Egg weight loss increased with each 2-day increase in storage. Egg storage did not significantly reduce hatchability when eggs were stored for a maximum of 6 days. However, after 8 days of storage, there was a significant decrease in of fertile eggs, and an increase in late period embryo mortality (between incubation days 15-21). Hatchability and late period embryo mortality again levelled off between 10 and 16 days of storage.

Table: *Egg weight loss during storage and incubation.*

Storage length (days)	*Fresh egg weight (g)*	*Post-storage egg wt. (g)*	*Loss during storage (g)*	*transfer egg wt (g)*
2	62.9	62.6	0.30	54.5
4	62.9	62.6	0.37	54.3
6	62.8	62.3	0.48	54.3
8	62.7	62.2	0.55	53.9
10	62.9	62.2	0.63	53.8
12	62.8	62.1	0.71	53.7
14	62.7	61.9	0.81	53.3
16	62.7	61.8	0.91	53.4

The Importance of Traying Eggs with the Large End Up

The egg possesses certain characteristics which promote the proper development of quality chicks. For instance, the shell permits respiration by allowing oxygen to pass through tiny pores in its surface. Until the chick is able to crack the shell, respiration can only occur with the help of oxygen that passes through these pores. In addition to the porous shell, the egg contains two membranes that also influence the chick's development.

These membranes closely line the inside of the shell and are loosely bound to the shell and to one another. The shell membrane closest to the shell is called the "outer shell membrane," and the one in contact with the albumen is called in the "inner shell membrane." During incubation, after the egg has been laid, an air cell will form at the large end as a result of the separation of the two membranes.

The embryo is normally oriented with its head in the large end of the shell. At day nineteen, the embryo will poke its beak between the separated membranes and use the air cell to breathe for the first time. The chick is allowed to "practice" breathing while remaining inside the shell, which allows for the maturity of different organs.

Effects of Placing Eggs Small End Up

In the event that an egg is placed small end up, the chances are greater that an embryo will situate its head in the small end of the egg. When an embryo is not situated correctly, reduced hatchability and lower chick quality are common problems that may occur. Sometimes a chick embryo is unable to locate the air cell and accidentally pips the small end of the shell, which causes the chick to hatch too early.

A recent study conducted by Bauer, Tullett, and Wilson found a significant difference between eggs set large end as opposed to small end up. When a commercial broiler strain was used, the eggs set small end up had a high percentage of disoriented embryos. Only 38.6% of the embryos situated themselves correctly and pipped the large end of the egg, while the remainder of the embryos became disoriented: 39.2% pipped the small end of the egg and 22.2% pipped at the equator. In addition, the researchers found that chick quality is lessened, and hatchability is reduced by approximately 17% when eggs are set small end up, as shown in Table 1. An increase in mortality occurred during last week of incubation—after day nineteen—when the embryo is normally poking its beak into the air cell, called "internal pipping."

Care in Preparing Egg Pack – A Necessity

Eggs may be placed upside down for several reasons, some of which include simple carelessness and the inability to determine which end is, in fact, the large end. Despite these difficulties, a concerted effort must be made to place the eggs large end up. Without this simple precaution, the egg cannot properly provide an environment for the developing embryo to hatch and become a high quality chick.

Specific Gravity Determination for Hatching Eggs

Specific gravity determinations are a non-invasive method to determine eggshell thickness, and therefore, eggshell quality. Specific gravity determinations also have the advantage of being simple, easy to perform, and inexpensive. Specific gravity reflects the amount of shell present relative to the amount of albumen, yolk, and membranes. It can be used by the poultry industry to assess shell quality of hatching eggs as the hen ages or during periods of stress which may affect shell quality. Specific gravity measurements have shown to decrease after approximately forty-five weeks of age. While it is expected that specific gravity will decrease with age of flock, sometimes the use of this procedure will allow one to know when specific gravity falls below an expected level. It will allow one to quantitate the shell quality of a flock at any particular age which may be having a problem and to also evaluate the effectiveness of a treatment used to improve shell quality. It has been shown that hatchability of thin shelled eggs (<1.080) was at least two percent less than thicker shelled eggs in hens of the same age and had a higher incidence of embryonic death.

The specific gravity of an object equals the weight of its volume relative to the weight of an equal volume of water, when both are at the same temperature. The specific gravity of an egg is equal to the egg's density relative to water. An egg has four basic parts: yolk, albumen, shell membranes, and shell. The specific gravity of all four parts of the egg are different (shell, 2.325; yolk, 1.032; albumen, 1.038; shell membranes, 1.075). Since the specific gravity of the shell is more than two times higher than the other parts of the egg, the percentage of egg that is shell has a major influence on the specific gravity of the whole egg. As the amount of shell increases, the specific gravity of the egg increases, and corresponding changes in the proportions of yolk, albumen, and shell membranes will be of little consequence. Egg specific gravity, therefore, is a good indicator of percentage shell and is the reason specific gravity is used to determine shell quality. At times a breeder manager or hatchery manager may wish to assess the status of egg shell quality for a particular flock or flocks.

To perform the specific gravity determination, you immerse the egg in a series of increasingly concentrated salt solutions until the egg floats on the surface of one of the solutions. Specific gravity is highly related to the incidence of breaks and cracks. The incidence of breakage is above normal if the specific gravity of a flock averages less than 1.080. It should be adequate that we use two to three salt solutions to give us an estimation of the specific gravity of a flock's eggs.

Equipment Required to Perform Specific Gravity Determinations

1. Three plastic 5 gallon containers with sealable lids, or 3 plastic dish pans
2. Feed grade or all-purpose non-iodised salt
3. One hydrometre with a specific gravity range of 1.050-1.100
4. One glass or clear plastic graduated cylinder for use with the hydrometre (must be taller than the hydrometre)

5. One perforated bucket or a plastic egg flat tot immerse the eggs in the salt solutions
6. Egg flats for each of the four levels of specific gravity determinations

The Specific Gravity Procedure

The day before the measurements are performed a minimum of 100 fresh eggs should be randomly selected from the flock. Avoid selecting only the better or poorer eggs from the flock. The eggs selected and the water to make the solutions should be placed in the room where the measurements will be performed the day before the test. This will allow all materials for the test to reach the same temperature overnight.

On the day of the determination, the salt solutions should be made. To make the salt solutions, add the corresponding amount of salt to the properly labelled plastic bucket containing 5 gallons of fresh, clean water (Table 1). The amounts given below are computed on the basis of one gallon, as some may prefer to use a dish pan to perform the test, which holds a lesser amount of water.

Solution Level	*Lbs. Salt / Gallon of Water*
1.070	0.8
1.075	0.9
1.080	1.0

You may wish to make a concentrated salt brine solution and then add appropriate amounts of water to arrive at the correct specific gravity solution. Be sure that the salt brine and the water are the same temperature.

These three solutions will yield four specific gravity groups: Less than 1.070, 1.070-1.075, 1.075-1.080, and greater than 1.080. Once the solutions have been made, they must be adjusted by using the hydrometre. The graduated cylinder is filled with a sample of each salt solution and the hydrometre is then placed in the graduated cylinder. The specific gravity of the solution should be read from the bottom of the meniscus for consistency.

If the reading is higher than desired, place the solution from the hydrometre back into the 5 gallon bucket and add a small amount of water. Stir thoroughly and test again. Repeat if necessary. Conversely, if the reading is lower than desired, return the solution from the hydrometre back into the 5 gallon bucket and add a small amount of salt. Stir thoroughly and repeat the test. Keep adding salt until you obtain the specific gravity you desire.

The perforated plastic bucket or the plastic egg flat may be used to lower the eggs into the solutions, starting with the lowest specific gravity solution and continue to the higher. If the egg floats in that solution, it should be removed to a flat correspondingly labelled to the appropriate specific gravity level. Table 2 demonstrates how to classify eggs into four different specific gravity groupings.

Since you are using a sample of 100 eggs, the percentage of eggs at each specific gravity level can be easily determined simply by counting the number of eggs in that specific gravity range. After the procedure is finished, the eggs may be rinsed with an egg sanitizer solution, allowed to dry, and used normally.

Specific Gravity of Solution	***Floating (F) or Sinking (S) Action of Egg***
1.070	*F* - less than 1.070
	S - greater than 1.070
1.075	*F* - between 1.070 and 1.075
	S - greater than 1.075
1.080	*F* - between 1.075 and 1.080
	S - greater than 1.080

Points to Remember

- Specific gravity measurements should be taken within 24 hours after the eggs are laid. Consistent timing of specific gravity determinations are important for reliable results which can be used to compare one trial with another. Moisture loss will reduce specific gravity measurements.
- The salt solutions can be sealed and reused but must be rechecked for accuracy prior to subsequent use. Adjustments will be necessary each time they are used.
- Specific gravity determinations should be performed the same way each time they are done. This permits an accurate comparison of scores over time, giving the breeder manager valuable information about the flock.
- The specific gravity readings will not be accurate if the temperatures of all materials have not been allowed to equilibrate overnight. Accurate specific gravity readings are very dependent on both the eggs and the solutions being at the same temperature during the test.
- Always obtain a random sample of eggs for the test. If you select out samples from only one area of the house, or if you use

consistently superior or inferior eggs, the results will not give you an accurate representation of the performance of your flock.

Vitamin Levels in Breeder Diets

Hubbard breeder hens are capable of peak egg production around 86% under commercial conditions, with sustained peaks over 80% for 12 weeks. This high egg output is only possible with superior management and nutrition, part of which is a well-fortified high energy diet. These high persistent peaks also mean that we have to supply adequate vitamins in the feed, not only for the hens very active stage of reproduction, but also to ensure optimum hatchability and chick quality. Unfortunately, what we consider to be optimum vitamin needs of the breeder are often questioned as being too high and too expensive, when in reality it is low-vitamin levels in the feed that are ultimately the "most-expensive" scenario for the bottom-line of the breeder or integrated broiler operation.

Diets for adult breeder hens probably contain the highest level of supplemental vitamins of any feed manufactured at a commercial feed mill. These vitamins are required for the normal function of most physiological processes in the body, and for the breeder hen we have the added need for adequate carry-over into the eggs. Unfortunately a deficiency of any one vitamin can cause a dramatic decline in egg production and hatchability, and yet the levels recommended by Hubbard and other breeders are often questioned as being too high, and therefore too costly. We usually meet the breeders' requirements for vitamins by adding all as synthetic sources. The regular feed ingredients, such as corn, soybean meal and meat meal, all contain "natural" sources of vitamins and in some situations could theoretically contain enough to meet the breeders' needs. However the concentration of vitamins contained in corn will, for example, be affected by seed variety, growing conditions, harvesting conditions and storage conditions. Likewise the vitamins in meat meal will vary greatly depending upon the animal components used and the time and temperature of cooking and drying during processing. Obviously these ingredients come with no guarantees of vitamin levels, and bioavailable concentrations are expected to vary 200-300%. In addition to this inherent variability is the effect that factors, such as natural plant toxins and mycotoxins, can have on vitamin availability. For example, recent research data showed that the effects of a certain mycotoxin could be largely overcome by adding more thaimin, a B-vitamin, to the diet. On this basis it is perhaps not too surprising that we do not rely on regular feed ingredients to supply vitamins, and consequently our vitamin premix is designed to supply all necessary vitamins for the bird. However this still does not answer the question of "why such high levels of vitamins in breeder feeds." The starting point used in developing vitamin requirements, is the NRC values which are updated each 6-8 years. These NRC values are absolute requirement values for individual vitamins, and most often reflect the level of vitamin needed to prevent deficiency symptoms. In feeding breeders we want not only to prevent signs of vitamin deficiency, but also to ensure good egg production, hatchability and early chick vitality. This superior performance will only be achieved by feeding much higher levels of vitamins as part of a balanced nutritional programme.

Another reason for higher vitamin fortifications relative to NRC (1994) is the loss in potency of vitamins that occur between feed manufacture and consumption by the bird. Different vitamins are susceptible to various stresses to varying degrees, but as a generalisation it can be stated that the major causes of loss of vitamin potency are storage time, storage temperature, and storage humidity of the premix before mixing, and of the feed after mixing. Another major loss of vitamins occurs if they are premixed with minerals and stored for any length of time prior to incorporation in feed. Also conditions within the premix and feed can cause loss of potency. For example, some vitamins are acidic whereas others break-down under acidic conditions. Finally to really cause problems to vitamin stability, we sometimes pellet feed, and here the temperature and humidity can cause vitamin breakdown. The manufactures of vitamins are able to tell us factors which affect potency of their vitamins, and so with this knowledge, coupled with anticipated field conditions, we can project necessary safety factors to ensure optimum breeder performance.

The next question is how much does it cost to add these higher levels of vitamins, and are they worth it? BASF, a major producer of pure vitamins, recently conducted a survey of breeder operations in the USA and, as anticipated, found a range of vitamin levels being used. This industry data covered 62 broiler companies and represents about 90% of the industry. The highest and lowest vitamin levels being used, and these are compared to NRC (1994). There is about a 100% difference between the top 25% and bottom 25% of reported levels for vitamin fortification. The cost of obtaining these individual vitamin levels, and the total cost per tonne of mixed feed.

Our recommendation is, therefore, to feed adequate levels in the feed, because this ensures optimum delivery of vitamin to the breeder and the developing embryo. With these extra levels of vitamins in the feed, there should be no need to use supplemental vitamins in the water, except in situations of environmental or disease stress when feed intake is not optimal, or there is evidence of enteritis.

If vitamin levels are to be adjusted according to local conditions, then which ones are most likely to influence feed cost? Table 2 is a repeat of the industry high-level breeder vitamin recommendations, but in this situation they are listed according to contribution to cost of the premix. This data shows vitamin E, biotin and vitamin A to be the most expensive, and together make up over 50% of the cost of the vitamin premix. These three vitamins should, therefore, receive most attention when additions to, or deletions from premixes, are contemplated based upon knowledge of specific needs within a feeding programme.

Table: *Vitamin recommendations ranked by $ cost/tonne feed.*

Vitamin	*$/tonne feed*	*% contribution to premix cost*
Vitamin E	1.08	23.7
Biotin	0.77	16.9
Vitamin A	0.68	14.9
Riboflavin	0.53	11.5
Pantothenic acid	0.38	8.3
Pyridoxine	0.31	6.8
Niacin	0.27	5.9
Vitamin K3	0.13	2.9
Folic acid	0.13	2.9
Thiamine	0.11	2.4
Vitamin B12	0.09	2.0
Vitamin D3	0.08	1.8
Total	$4.56	100.0

Vitamins are expensive, and together the premix represents about 2-3% of the cost of breeder feed. However the effects of deficient or marginal levels in breeder feed can be far greater than the 10? per breeder to be potentially saved by feeding at the low vs high end of current industry standards.

Lysine Requirements of Pre-lay Broiler Breeder Pullets: Determination by Indicator Amino Acid Oxidation[1,2]

Materials and Methods

The University of Alberta Faculty Animal Policy and Welfare Committee approved all experimental procedures. Female Cobb 500 broiler breeder pullets (17 wk of age, n = 4) were selected from a commercial farm and transferred to the Metabolic Research Unit at the University of Alberta. The pullets were caged individually; they were given 2 wk to adapt to their new environment and 1 wk to recover from surgery in which a catheter was implanted into the left jugular vein. An 8-h:16-h light:dark schedule was used. The lights were automatically turned on at 0800 h each day. The birds were fed 75 g/d of test diets based on management of BW according to the Cobb 500 breeder management guide during the 3-wk trial. For the first 14 d of the study, the birds were fed a standard mash broiler breeder pullet diet formulated to meet NRC recommendations. The diet was then changed to a control (9.6 g lysine/kg diet) pelleted, purified diet, which was formulated to meet the estimated lysine requirement of broiler breeder hens at peak production according to the NRC on the basis of BW with a calculated metabolisable energy (ME) of 15.7 MJ. The amino acid mixture was interpolated (using recommendations from D. H. Baker, University of Illinois, Urbana, IL, personal communication) from requirements for 42- to 56-d-old broiler chicks fed purified diets containing 133 g/kg diet protein equivalent and 14.9 MJ ME/kg diet (23) and requirements for broiler breeder hens in peak production (18). The diet was formulated to ensure that all indispensable AA were in excess of the NRC requirement for broiler breeder layers (18) by at least 10% except for the AA_{test} (lysine) to ensure that lysine was the only limiting AA during the experiment. This diet was fed to the pullets 3 d before surgery and continued until the first oxidation study at 20 wk of age.

The composition of pelleted, purified test diets (g/kg diet) was: cornstarch, 624.8; corn oil, 100.0; salt mix,[5] 53.7; cellulose (Solka Floc; International Fibre, North Tonawanda, NY), 20.9; vitamin mix (AIN-93-VX; 960402, ICN Biomedicals, Aurora, OH), 10.0; all diets contained 189.4 g of AA mixture.[6] In addition, three basal diets were formulated, each having different levels of L-lysine-HCl (0, 4.8 and 14.4 g/kg diet) balanced with L-glutamate at 134.4, 129.6 and 120.0 g/kg diet, respectively, to keep each of the diets isonitrogenous. Each of the nine AA_{test} diets was obtained by mixing together the correct ratio of a basal diet above and below the AA_{test} level.

Twelve hours before surgery at 19 wk of age, feed and water were removed from the birds. The surgical insertion of a catheter into the left jugular vein was described. Observation from previous experiments (13) indicated that a minimum of 3 d was required for birds to recover from surgery, as indicated by daily consumption of a quantity of diet equal to the presurgery amount, activity level and responsiveness to the environment. Beginning at 20 wk of age, at <“0900 h on each oxidation day, the birds were weighed and placed into individual metabolic chambers for determination of oxidation rate. Initially, each bird was fed the control diet (9.6 g lysine/kg of diet) and oxidation of the indicator amino acid was determined. Subsequently, each bird received one of the eight additional test levels of lysine for an adaptation period of 43 h before each oxidation study. Assignment of dietary treatments to each bird was in random order until each bird had been fed each test level of lysine. Basal diets were mixed proportionally to obtain the required lysine concentration.

Tracer infusion and $^{14}CO_2$ recovery

Each of the test diets was fed for 43 h; the oxidation rate over a 4-h period was determined using the IAAO method. L-[1-^{14}C]phenylalanine (American Radiolabelled Chemicals, St. Louis, MO), prepared in a sterile saline solution with a concentration of 37 kBq/mL, was used as the indicator AA. Each bird received a priming dose of 74 kBq/kg body followed by a 4-h constant infusion of 44 kBq $\cdot$ h^{-1} $\cdot$ kg $\cdot$ body^{-1} via the jugular catheter using a syringe pump (KDS 100, Fisher Scientific, Mississauga, Canada). Half of the daily feed allocation of 75 g was divided into four equal portions and fed to the chickens hourly beginning immediately after the initiation of the continuous infusion. The feeding protocol was designed to minimise any change in total breath CO_2 excretion due to feed intake over the 4-h breath collection period. After the completion of the oxidation experiment, the birds were transferred back into individual cages. The next randomly selected test diet was offered to the birds, comprising the remaining daily feed allocation of 37.5 g.

Total breath was collected and sampled every 30 min during the constant infusion period, as described previously. For each oxidation study on an individual bird, the ^{14}C-phenylalanine oxidised and recovered from breath as $^{14}CO_2$ (% of dose infused) was plotted against time. To accurately determine the percentage of dose oxidised, the oxidation rate must be at steady state, as indicated by a plateau in oxidation rate (no change in oxidation rate over time). Steady state was first determined by visual inspection and confirmed by nonsignificant linear regression. The rate of $^{14}CO_2$ expiration was then corrected for retention of the label in the bicarbonate pool.

Statistical analysis

This study was based on repeated measurements within animals, in which each bird received each dietary treatment in random order.

Dietary lysine concentration was the main treatment effect. General linear models using the repeated-measures option determined differences among treatments. Significance was taken at a level of $P < 0.05$; differences between least-square means were obtained by pair-wise comparisons using Tukey's test as an option within the general linear model procedure. Lysine requirements were established using a linear-plateau model adapted from SAS, regressing oxidation rate on dietary lysine concentration. The breakpoint between the slope and plateau phases of the model was changed iteratively until the residual mean-square error was minimised. The individual requirements were used to determine the population mean and its SD. An upper 95% CI for the breakpoint was calculated from the mean requirements.

RESULTS

The birds were in good physical condition, with no signs of stress (inspected at least twice per day) during the oxidation period. Mean BW gain (±SD) during the 17-d experimental period was 202 ± 26 g. The Cobb 500 management guide recommends 340 g BW gains over the same period. The restricted but steady growth of birds was assumed to be due to the limited lysine in some of the test diets fed to the birds during portions of the experimental period. Two observations (Bird 2, 7.68 g lysine/kg diet and Bird 4, 0.96 g lysine/kg diet) were removed from the data set due to technical difficulties during the infusion period. The typical response pattern seen in IAAO of $^{14}CO_2$ recovered over time with bird 1 fed nine levels of lysine/kg of diet. The response pattern falls as the AA_{test} intake is increased until the requirement is reached. At that stage, there is no further response to additional AA_{test} intake.

The mean plateau CV for all oxidations was 8.93 ± 4.32%/kg body. Change in BW at each oxidation was taken into account. The mean percentage of ^{14}C-phenylalanine oxidised decreased linearly from the 0.48 to 4.80 g lysine/kg diet treatments and remained constant from the 4.80 to 14.4 g lysine/kg diet treatments. The lysine requirement for the individual bird was determined by a one-slope broken-line model. The number of observations with the least-square means value, which included the 4.80 g lysine/kg diet, determined the slope of the line below the estimated lysine requirement. The individual lysine requirements (breakpoint) for Birds 1, 2, 3 and 4 were determined to be 4.6, 3.9, 4.8 and 6.2 g lysine/kg diet or 345, 293, 360, 465 mg lysine $\cdot$ hen^{-1} $\cdot$ d^{-1}, respectively. The mean requirement was 4.88 ± 0.96 g lysine/kg diet or 365.6 ± 62.6 mg lysine $\cdot$ hen^{-1} $\cdot$ d^{-1}. The lysine intake necessary to meet the requirement of 95% of the population (upper 95% CI) 20–23 wk of age was calculated as 6.4 g lysine/kg diet or 480 mg lysine $\cdot$ hen^{-1} $\cdot$ d^{-1}.

Discussion

The difficulty in adapting traditional AA requirement methodologies has resulted in a scarcity of scientific data regarding the requirements of broiler breeders. Currently, the NRC (18) provides nutrient recommendations only for mature broiler breeder hens at the peak of egg production. Of those recommendations, many are not derived from experimental data, but rather from interpolations based on the requirements of laying hens and broiler chickens (18). These interpolations are necessary, given the lack of scientific data, but are at best educated guesses of the true nutrient requirements. The growing broiler chicken, although genetically similar to the broiler breeder pullet, is at an entirely different physiologic stage; the laying hen has been genetically selected for very different production traits than the broiler breeder. By using the IAAO method in this study, we demonstrated that the lysine requirement of broiler breeder pullets is substantially lower than that recommended for mature broiler breeder hens at peak egg production [765 mg lysine $\cdot$ hen^{-1} $\cdot$ d^{-1}; (18)]. These data are not surprising because pullets have a relatively low growth rate, are not sexually mature and do not have a demand for egg formation.

Fisher described an estimate of the lysine requirement of broiler-breeder hens at peak production as a function of maintenance, growth, egg production and flock variability. The birds studied by Fisher were of similar age and BW as the birds used in our study (mean BW 2333 ± 246 g), except that our birds had not been induced to lay through photostimulation. Therefore, it is reasonable to assume that the lysine requirement of our birds would be equivalent to the estimate by Fisher of 893 mg lysine $\cdot$ hen^{-1} $\cdot$ d^{-1}; minus the requirement for egg production. Fisher estimated that 50% of the lysine requirement at peak production was due to egg production, so that the lysine requirement for the other functions (i.e., growth and maintenance) would be 447 mg lysine $\cdot$ hen^{-1} $\cdot$ d^{-1}. Applying the same estimated partitioning to the lysine requirement for broiler-breeder hens according to the NRC, [765 mg lysine $\cdot$ hen^{-1} $\cdot$ d^{-1}], the requirement minus egg production would be 383 mg lysine $\cdot$ hen^{-1} $\cdot$ d^{-1}. In our study, we determined mean a lysine requirement of 365 mg lysine $\cdot$ hen^{-1} $\cdot$ d^{-1}, with an upper 95% CI of 480 mg lysine $\cdot$ hen^{-1} $\cdot$ d^{-1}. Our mean requirement agrees with the predicted nonlaying lysine requirement extrapolated from the values published by the NRC.

Moreover, our calculated "safe intake" or upper 95% CI just exceeds the requirements calculated from Fisher. It is important to note that this nonlaying requirement includes the lysine needed for growth as well as maintenance. Although our birds grew during the 3 wk of experimentation (increasing their BW by <“10%), the frequent periods of lysine deficiency and the necessary feed restriction make the lysine requirement for growth and for maintenance impossible to calculate. However, IAAO allows the calculation of total metabolic requirement for lysine. In addition, previous estimates for the lysine requirement for maintenance alone are highly variable and have been estimated at 13–215 mg lysine $\cdot hen^{-1} \cdot d^{-1}$ (23,25). In conclusion, our results for the lysine requirement of prelaying broiler breeder pullets are confirmed by similar estimates calculated from existing literature data. The IAAO method is sensitive to rapid changes in the AA pool (intracellular protein synthesis) within the test animal and does not rely on indirect responses such as growth and egg production to determine AA requirements. This principle makes the IAAO an excellent method with which to determine essential or conditionally indispensable AA requirements in broiler breeder pullets.

The results generated using this method are ideally suited to inclusion in mathematical models of poultry AA metabolism and requirements for maintenance, egg production and growth, which Fisher suggested to obtain more accurate determinations of requirements in poultry. The IAAO technique also allows for the determination of the amino acid requirements of individual birds over a very short period of time. This capability allows, for the first time, an accurate assessment of the variability of requirements within a population, which would allow the development of stochastic models for growth and requirement.

In addition, these flock variability data will allow nutritionists to determine accurately the cost of meeting the lysine requirements of a greater proportion of individual birds. To account for population variance in requirement, feeding at the 95% CI (6.4 g lysine/kg diet) will meet the requirement of 95% of the birds. This safety margin for feed formulations will have considerable economic effect with respect to efficiency of production. Furthermore, these results provide poultry nutritionists with a starting point from which to calculate the ideal AA requirements for broiler breeder pullets (20–23 wk of age) using lysine as the reference AA. Swine nutritionists have used ideal AA requirements for a number of years; this approach has gained acceptance in the poultry industry as well. Factorial modelling has been used to determine energy and AA requirements in poultry. Data collected using the IAAO method have the potential to make calculations based on factorial modelling more accurate.

Until now there has not been a suitable method with which to determine AA requirements in individual birds. The IAAO method was adapted to determine individual breeder pullet requirements because it allows multiple measurements to be performed within the same animal, without a change in physiologic state during the experiment. The determination of the lysine requirement of individual breeder pullets allows nutritionists to make an accurate assessment of the cost of meeting the lysine requirements of an increased proportion of the population. The AA requirements of broiler breeders would be expected to change substantially as they increase in frame size early in life, as they approach sexual maturity (as in the present study), as they reach sexual maturity and peak egg production and as egg production declines with age. The IAAO methodology will enable the determination of individual bird requirements at each of these stages, thus allowing a much more precise formulation of broiler breeder diets.

The Chicken Tractor

A chicken tractor, or chicken ark if you are in merry ole England, is a great alternative when it comes to keeping chickens. It is portable and can be moved around your yard. This portability can be very beneficial as it can provide fertilizer in different parts of your yard, keep the birds from wearing down one area, and it can help the birds find fresh bugs to eat and grass to graze. You can find many tractors for sale online, or you can choose to build your own. If you have common tools and a little know how, you can easily build your own. We recently built a couple of tractors to try them out ourselves. Below you will find a picture by picture guide to how we built ours.

What is a Broody Hen?

A broody hen is either a wonderful or a terrible thing depending on who you ask. If you're interested in breeding your chickens, a broody hen or setting hen as they are sometimes called is the most natural way to incubate eggs. If you just want eggs and don't have any fertile ones, a setting hen can be irritating. You see, a hen in this state is going to sit on her eggs regardless of fertility and she's going to be extremely unhappy if you try to move her or take them away from her. She will also quit laying for at least 2 months, possibly more, if allowed to set. One of the big questions among people who raise chickens is whether or not to let a broody hen stay that way. If you want to get the most egg production out of your hens, disrupting her broodiness, if possible, is a good idea.

Heritage or non-Production breeds are the most likely to go broody. Production bred hens are just that, bred for the most egg production. These have a short healthy life and commercially are replaced after their first and best year of laying then used for meat.If you want to hatch a batch of chicks you may wish to let her go through the cycle which takes about twenty-one days. You can also prevent broodiness in the first place as long as you know what you're doing. To keep a hen from setting, you'll need to avoid letting eggs collect in a nest. This collection of eggs triggers broodiness. If it's too late to do this, you can move her from the nest, put her in her own coop, or cover the nest so she can't get in. You can even use a "broody coop" – a hanging cage made from wire that will house her until she gives up sitting on the eggs (this helps cool-off her underside which breaks the broody hormone cycle). The longer you take to break her of her broodiness the longer it will take her to lay again. Of course, if you're breeding chickens for meat or for fun, you can collect the eggs from all your birds and let the hen sit on as many as she can handle at once. This will depend on the size of the hen and the size of the eggs. 8-12 eggs in a clutch is the average. She'll incubate them just as effectively if not more than an incubator with a bonus; you'll need no brooder to keep the chicks warm until they are fully feathered and nearly half grown, if she is a good mother (not all hens are). Natural incubation is usually only effective in the spring and summer, when hens go broody, so hatching at other times of year will require artificial incubation. Using a broody hen that wants to be a mother can be an excellent way to raise baby chicks. For best results the hen and chicks will need to be housed separately from the rest while the babies are growing up. There are exceptions to every rule and many flocks with a large yard and safe place for chicks and hen on the ground at night naturally accept and don't bother chicks being raised by a hen. Chicks raised away from the flock often shouldn't be introduced to the flock until they are near adult size. This can take a few months depending on the breed's growth rate. Some people hatch the chicks under a broody hen, then raise them separately. Either way, make sure that your hen has plenty of food and water - some may not move off the nest to eat or drink for weeks, especially in cool weather.

Taking a setting hen off the nest once a day, offering food and water, will help her stay healthy through the 21 day incubation process. Most utility or Production chickens - breeds that produce mostly eggs - have been bred not to become broody. Heavy, dual purpose, or bantam chickens are usually the best mother hens. Knowing which breeds are likely to become broody will let you avoid a setting hen if you want mostly eggs, or choose one just for this trait if you want to produce chicks. Whether a broody hen is a great thing or an awful one depends on what you want to do with your chickens.

Chapter 3

Breeder Nutrition

The developing embryo and the hatched chick are completely dependent for their growth and development on nutrients deposited in the egg. Consequently the physiological status of the chick at hatching is greatly influenced by the nutrition of the breeder hen which will influence chick size, vigour and the immune status of the chick.

Table: *The necessary change in hatchery or broiler performance to equalise profitability when breeder feed cost is changed by 1% per tonne (for example from £UK 140.00/tonne to £UK141.40/tonne or £UK138.60/tonne).*

Hatch of total eggs (%)	0.24
42 day liveweight (g)	7.4
42 day FCR	0.0015
42 day mortality (%)	0.07-0.45*

**depending on age of mortality. Calculated using input-output values for UK industry 2003 (Kemp and Kenny 2003).*

The Financial Effects

Nutritional decisions for breeders need to take account of the overall economics of the whole production cycle. Table shows the changes in hatchery and broiler performance that are required to equalise the effect of a 1% increase in breeder feed cost on the profitability of the whole production cycle. Only one of these changes is required to have the necessary economic effect; in practice all are likely to move positively making the measurements of any one change difficult.

The calculations are done under typical UK 2003 conditions and they show quite clearly that small improvements in bird performance are required to 'pay' for more expensive breeder feed. Conversely, apparent savings in breeder feed cost can readily lead to an overall loss if small changes in broiler performances are ignored.

Similar economic analyses have been conducted by Mississippi State University which, based on US integration 2002 costs, demonstrates that a measurable improvement in progeny liveability as a result of hen diet change can be profitable. The key point is that trying to cut the cost of a breeder feed may easily reduce the profitability of the overall enterprise.

Influence of Feed Allocation

Underfeeding the hen can have an impact on chick quality and this is particularly noticeable in the early production period. Modern hybrid parent flocks commence production at a faster rate than in the past and consequently egg output increases over a shorter time span during the early laying period. Feed allocations during this period have not necessarily increased in line with this egg production trend. Low feed allocation intake by young commercial breeder flocks has been shown to compromise nutrient transfer to the egg, resulting in increased late embryonic death, poorer chick viability and uniformity.

In a recent study by Leeson (2004) broiler breeders were fed different levels of feed through peak production varying from 140 to 175 grams. Although the increased feed allocation increased bodyweight there was no influence on egg size, however chick weight was influenced by feed allocation. Of equal importance is the effect of overfeeding on ovarian development. In experimental studies ad libitum feeding has been the most widely used model for overfeeding which can result in excessive follicular development or Erratic Oviposition and Defective Egg Syndrome (EODES).

Flocks with EODES generally have poor shell quality, a reduced duration of fertility and poor hatchability. It is also known that fewer sperm will survive but it is not clear how the surviving sperm are affected and if they generate a weaker embryo. The same authors also warn that the effect of aggressive feeding two to four weeks after photostimulation reduces productive performance throughout the life of the flock.In this period the bird switches from primarily growth to a reproductive state. The young birds' reproductive hormone system is not mature enough to deal with high nutrient intakes; nutrients are instead metabolised to egg yolk lipid which contributes to excess follicle development.

Research shows that nutrient supply to the broiler breeder is of consequence to chick quality and production performance. This places greater emphasis on the nutritionist providing the correct nutrient density diet and the flock manager to provide appropriate feed intake to the bird coming into lay.

Table: *The effects of breeder feed levels on chick weight.*

Peak breeder feed (g/b/d)	30 week breeder chick weight (g)
140	40.3
147	40.0
155	41.5
162	41.7
169	41.8
175	42.0

Diluted Breeder Diets

The use of diluted breeder diets is receiving a lot of attention in Western Europe on the basis of improvements in bird welfare. Experimental work feeding low energy density diets to young parent stock gave a delayed onset of oviduct development, increased early egg size, faster development of the embryo and a higher live weight of day old chicks. When broiler mortality was above average, low density broiler breeder feeds gave a significant reduction in mortality of offspring. Other experimental work showed improvements in breeder productive performance when diluted diets were fed in the rearing period.

Vitamins

Vitamins are involved in most metabolic processes and are an integral part of foetal development, therefore the consequence of suboptimal levels of these nutrients in commercial diets are known to result in negative responses to both parent and offspring performance. Vitamins account for about 4% of the cost of a breeder feed, so economising on vitamin inclusion rates is rarely an option

Table: *Some practical recommendations for vitamin supplementation of breeder feeds*

Vitamin	Leeson & Summers (1997)	DSM	Ross
A (iu/g)	7	10-14	13
D3 (iu/g)	3	2.5-3.0	3
E (mg/kg)	25	40-80	100
K (mg/kg)	3	2-4	5
Thiamine (mg/kg)	2.2	2-3	3

Contd...

Vitamin	*Leeson & Summers (1997)*	*DSM*	*Ross*
Riboflavin (mg/kg)	10	8-12	12
Pyridoxine (mg/kg)	2.5	4-6	6
B12 (mg/kg)	0.013	0.02-0.04	0.03
Nicotinic acid (mg/kg)	40	30-60	50
D-pantothenic acid (mg/kg)	14	12-15	12
Biotin (mg/kg)	0.2	0.2-0.4	0.3
Folic acid (mg/kg)	1	1.5-2.5	2

Generally there is a shortage of information on vitamin requirements of broiler breeders especially when related to offspring performance.

Most of the breeder work is quite dated and since that time breeder performance has changed. It would be impossible to review all the literature in this article, however a review of work on fat soluble vitamins, biotin and pantothenic acid have shown that vitamin E has the largest impact on progeny.

Table: *Impact of dietary breeder vitamin status on bodyweight, enzyme activities, tissue characteristics and immunity of progeny.*

Vitamin	*Progeny response*
Vitamin A	Increased liver vitamin A in embryonic and chick liver but decreased vitamin E, carotenoids and ascorbic acid. Surai et al. (1998).
Carotenoids	No positive impact on chick growth, organ development or humoral immunity in chicks five weeks post-hatching. Haq et al. (1995).
Carotenoids	Transferred from the hen to the yolk but not absorbed well by the embryo and subsequent chick. Haq and Bailey (1996).
Carotenoids and Vitamin E	Carotene, vitamin E, and their combination improved and vitamin E lymphocyte proliferation, but only vitamin E improved humoral immunity. Haq et al. (1996).
Vitamin E	Vitamin E levels of 150 and 450mg/kg increased passively transferred antibody levels in chicks to Brucella abortus up to seven days of age. Jackson et al. (1978).
Vitamin E	Increased vitamin E in chicks' yolk sac membrane, liver, brain and lung all of which had reduced susceptibility to peroxidation. Surai et al. (1999).

Contd...

Vitamin	***Progeny response***
Vitamin E	Increased progeny antibody titers to sheep red blood cells at hatch. Boa-Amponsem et al. (2001).
Vitamin E and Selenium	Increased liver glutathione activity in chicks. Increasing selenium increased selenium dependant glutathione peroxidase in chick liver. Surai (2000).
Vitamin D	Tibial calcium was increased at two weeks post-hatching and tibial ash increased at four weeks of age by increased vitamin D3. Ameenudin et al. (1986).
Vitamin K	Chicks from hens fed vitamin K deficient diet had reduced tibial glutamic acid levels at day one and 28 post-hatching but tibial glutamic acid was restored by supplementing the chick diet with vitamin K. Lavelle et al. (1994).
Biotin	Foot pad dermatitis and incidence of breast blisters were decreased in some trials in chicks from hens fed biotin fortified diet. Harms et al. (1976).
Biotin	As biotin increased in the hens' diet, yolk and chick plasma also increased. Biotin concentration in chick plasma was poorest from young hens. Whitehead (1984).
Pantothenic acid	Liveability of chicks was best when hens were fed 20mg/kg diet of pantothenic acid. Utno and Klieste (1971).

The production and economic effects of vitamin E supplementation are best shown by Hossain et al (1998) where a basal corn soya feed was supplemented with 25, 50, 75 and 100mg/kg vitamin E. The effects on hatchability were not significant; however the best hatchability was obtained at 50mg/kg at 52 weeks. Offspring immune response continued to increase up to 100mg/kg. In the same studies higher final bodyweights at 42 days, improved FCR and reduced mortality were observed in chicks from eggs which had been injected with vitamin E in ovo.

Haq et al., (1996) working with very high levels of vitamin E (134mg/kg versus 412mg/kg) found no growth response to 21 days and an improvement in FCR for the offspring of hens receiving the supplemental feed. In other studies the combination of selenium and vitamin E to broiler breeders has been shown to increase liver glutathione activity of progeny. In general it seems to be justified to supplement practical breeder feeds with 100mg/kg vitamin E.

There appear to be mixed reports on the efficacy of vitamin C; some experiments suggest a positive response, but a more recent study failed to detect any benefit on any production parameter.

This lengthy study used corn soya diets supplemented with 75mg/kg stabilised vitamin C which when analysed recovered 49mg/kg which might explain the variability of response. The influence of increased vitamin levels fed to young parent stock on progeny performance is an area which has received significant commercial interest.

Work conducted at Aviagen Ltd has shown chicks derived from 31 week old parent stock fed elevated levels of vitamins showed improved growth to 11 days and reduced mortality compared to chicks derived from 42 and 45 week old parents. Similar responses have been found in the field where chicks derived from young parents fed increased levels of vitamins have benefited in terms of viability and liveability. Perhaps this supports the need for further work exploring the vitamin requirements of the breeder in the early production period.

Table: *Blood cell count of the broiler derived from parents fed high or low vitamin and mineral levels*

	Breeder low vitamins/minerals	*Breeder high vitamins/minerals*
Heterophil	5.3	3.8
Lymphocyte	4.6^{a}	21.4^{b}
Monocyte	1.1	5.3
Basophil	0.0^{a}	5.4^{b}

Whitehead (1991) proposes that a basis for making recommendations is to feed vitamin levels that maximise the resulting level in the egg. For vitamins with active transport mechanisms (thiamine, riboflavin, biotin, cobalamin, retinol and cholecalciferol) these levels reflect the saturation of binding proteins. Levels derived in this way include 10mg/kg for riboflavin and 250 microgram/ kg for biotin. Whitehead (1991) contrasts this level of riboflavin with the conventional requirement (4mg/kg in this case) but the higher figure – the upper limit to nutritionally useful range – may be a better guide to good commercial practice.

Table: *Summary of minerals fed to breeders shown to have an effect on progeny performance.*

	Growth	***Liveability***	***Immune function***	***Skeletal***
Fluoride				X
Phosphorus				X
Selenium		X		
Selenomethionine	X		X	
Zinc	X		X	X
Zinc and methionine		X		

Vitamins and Chick Immunity

Reference has already been made to the effect of vitamin E on chick health and immune function, while other vitamins have been researched none show the same degree of effect as vitamin E. Recent work by Rebel et al (2004) investigated the effects of several elevated levels of vitamins and trace elements fed to breeders and broilers on the immune system of birds infected with malabsorption syndrome. Broilers derived from breeders fed elevated vitamins and mineral levels had increased numbers of leukocytes at day old which indicated stimulation of the immune system.

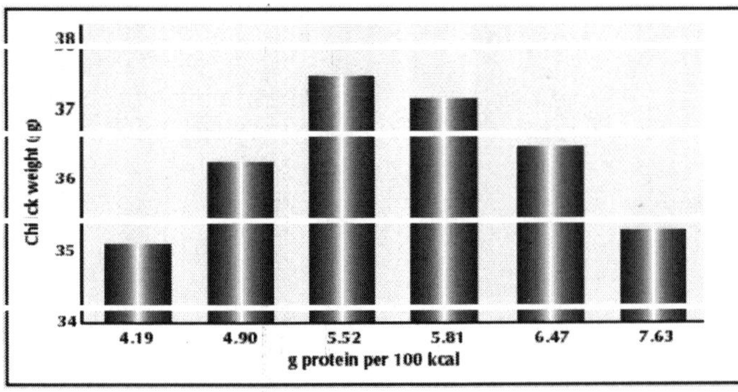

Figure: *The effect of protein-to-energy ratio in the breeder feed on chick weight at hatch*

Major Minerals

Calcium, phosphorus, sodium, potassium, magnesium and chloride are involved in shell formation hence general improvements in shell quality lead to better egg and chick quality. Variations in maternal phosphorus supply have been shown to influence bone ash of young but not older progeny. Broiler performance was not affected by these treatments so the practical significance of this work is not clear but the use of relatively low phosphorus levels in breeder diets, while benefiting egg shell quality, may not lead to the best possible bone integrity in the early stages of growth.

Trace Minerals

Most interest in this field has centred on the use of chelated minerals which have been shown to increase deposition in the egg and transfer to the tissues of the hen and the embryo. Most recent work has focused on the antioxidant status of breeders, embryos, offspring and the role of selenium. Surai (2000) has shown the role of Selenomethionine on both vitamin E and glutathione peroxidase levels in eggs, embryos and chicks up to 10 days of age. The economic benefits of using Selenomethionine compared with sodium selenite have been examined in a number of unpublished field trials in the UK. Hatchability improvements ranged between 0.5-2.0 chicks per 100 eggs and in another trial 0.3-0.7 chicks per 100 fertile eggs. Few of these tests involve a proper assessment of subsequent broiler performance although comments about chick quality are generally positive.

In one of the commercial trials mentioned an improvement of 0.5% in mortality and cull rate at 10 days was observed when organic selenium replaced sodium selenite. Research has indicated that the improvements in chick immunity as a result of mineral fortification of hen diets may result in improved liveability. Flinchum et al. (1989) demonstrated that leghorn breeders fed supplemental zinc methionine to a zinc adequate diet had progeny with improved survival to an E. coli challenge. Similar improvements to progeny liveability were seen with breeders fed supplemental zinc and manganese amino acid complexes.

Nutrient Levels in the Breeder Diet

There is clear evidence that a high protein to energy ratio depresses hatchability, and probably chick performance. The experiment by Whitehead et al. (1985) shows the effect of excess protein where the higher protein level reduced reproductive performance, producing 3.1 fewer chicks per 100 fertile eggs. Chick quality was also reduced so that the difference in saleable chicks was 4.0 per 100 fertile eggs. The effect of energy protein ratio in the breeder feed is shown in figure. This emphasises both the effects of excess and inadequate protein, and also indicates that the optimum level is quite steeply defined. According to this trial the optimum protein level is at 5.52g protein per 100kcal which converts to an optimum of 15.18% protein for a diet containing 2,750cal/kg of feed. The protein level of the diet and its ratio to energy is important not only for parent performance but also for chick quality.

Table: *Commercial comparison of breeder feeds based on wheat or maize (400g/kg)*

	Advantage of maize over wheat based feed
Mortality during lay (%)	-1.7
Total eggs (per hen housed)	+3.8
Hatching eggs (per hen housed)	+4.8
Hatching/total eggs (%)	+0.9
Hatch of set eggs (%)	+0.6
Hatch of fertile eggs (%)	+1.1
Second quality chicks	-0.1

The Effect of Feed Ingredients

There is evidence of improved breeder performance when maize is compared to wheat as the main cereal in breeder feeds. From a survey of many depleted commercial flocks overall hatch of fertile eggs in the UK based on wheat diets and Brazil based on maize diets is 83.3 and 86.2 per 100 eggs respectively. Other management factors may contribute to this difference in hatchability other than cereal source; male management is very good in Brazil and the resulting high fertility may also contribute something to this difference. Unpublished commercial development trials from the Netherlands and Aviagen Ltd grandparent flocks support this observation.

The most likely benefit of maize is probably in shell quality and thickness. From the same data average poorer shells with specific gravity of <1.08 accounted for 26.1% of eggs from wheat fed hens and 17.1% from maize fed. Studies of hatching losses showed less late dead embryos (>18 days) and less bacterial contamination.

These two responses are expected with eggs of better shell quality. Evidence about fat levels and sources is conflicting but there is no question that this is an important consideration. Added fat levels should be kept low in breeder feed (1-3%) and preference given to unsaturated vegetable oils rather than saturated animal fats.

Work from Mississippi State University compared maize oil and poultry fat and generally supported the use of more unsaturated fat.

Maize oil increased 21 day bodyweight over that of poultry fat and improved slaughter weight of broilers in comparison to equal levels of poultry fat and lard.

Table: *Experiments comparing fat sources and/or levels for broiler breeders.*

Reference	***Fats compared***
Brake (1990)	PF
Brake et al. (1989)	PF
Denbow & Hulet (1995)	SBO, PF, FO
Peebles et al. (1999a, b)	CO, PF, LA
Peebles et al. (2000a)	PF, CO, LA
Peebles et al. (2000b)	PF, CO, LA

Fats: PF – poultry fat; SBO – soybean oil; FO – fish oil; CO – corn oil; LA – lard

Over and undersupply of nutrients into and through lay can have a very significant impact on breeder production and quality of progeny.

This places greater emphasis on the nutritionist providing the correct nutrient density diet and the flock manager to provide appropriate feed allocation in lay.

Addition of micronutrients to the breeder has been shown to be beneficial to progeny quality especially in the early production period. Use of specific dietary ingredients such as maize can affect breeder performance and progeny quality. Both on economic grounds and on biological grounds, high quality nutrition of breeders is well justified.

Broiler Breeders - Feeding Breeders to Optimise Chick Quality

Integration concentrates the majority of their energy and technical resources on broiler production. Breeders generally do not receive the attention they deserve. Integrators are looking to maximise egg and chick production at the lowest cost with insufficient attention being given to the quality of the chick. There is an abundance of evidence that confirms that breeder nutrition directly impacts the progeny. However, these impacts are generally difficult to measure under commercial conditions and even under controlled trial conditions they may be obscured.

The obsession with lowering feed cost has, I believe, led us to a point where some nutrients, like vitamins and trace minerals are marginal in modern breeders. While, others like crude protein are generally fed in considerable excess and this profoundly affects breeder performance and health.

Due to fierce competition and the subsequent low profit margins, vitamin producers are reluctant to invest resources in research, so little breeder vitamin and trace mineral research has been conducted over the past 15 to 20 years. In addition, the current benchmarking practices have pressured the industry in lowering cost wherever possible.

Fifteen years ago, breeders were generally peak fed 165 to 170 grams per bird per day with peak egg production at 65 to 70 per cent. Today, the majority of breeders receive 145 to 150 grams of feed at peak with peak production at 80 to 90 per cent – this on reduced vitamin and trace mineral levels.

Measuring breeder nutrition impact on progeny is difficult due to a number of factors. Firstly, the assumption is that all birds receive the allocated feed amount; this certainly is not the case with 30- to 40-gram differences between birds being common. The birds that will be nutrient-limiting are those that are the most severely restricted. This may only be five per cent of the flock, so when assessing the commercial performance between two nutritional regimens it is difficult to have a large enough sample to be able to measure the impact. Secondly, some of the effects may only be apparent under stress conditions. Small or no differences may be seen under trial or ideal commercial conditions, but under a stress (disease, low broiler nutrient levels, chilling), the responses will be more apparent.

The Nutritionist's Role in Breeder Nutrition

Since breeders are restrict fed, it is the nutritionists responsibility to design a feed that meets the nutrient requirements of the bird under their particular set of conditions of; egg production, environmental temperature, etc. and then to ensure that the highest quality ingredients are used in these feeds and that the feeds remain constant over time. Since the birds are restrict fed, the breeder manager needs to 'learn' how to feed the birds. With constantly changing feeds this makes every flock a new experiment, which makes it difficult to manage body weights and uniformity.

Birds do not understand percentages. Producers focus on the nutrient composition of the feed. The most common question I hear is, "What crude protein and energy level should I be feeding?" We need to understand the daily nutrient needs of the bird for a particular environment, given amount of activity and egg production. We then need to know what feeding system is being used and how well feed is distributed to the birds. Once we have this information we know daily need for: kcals of ME, mg of amino acid, grams of calcium etc that the bird requires and how many grams we need to feed it. Then we can formulate the feed to meet these requirements.

Energy and Fat

Energy is a fuel for maintaining basic metabolism, mobility and egg production. Energy is not a nutrient as such but is the result of metabolic oxidation of energy yielding nutrients – carbohydrates, fats and proteins. Having reviewed a large number of breeder formulations, energy appears to be the fist limiting 'nutrient'. This is true under both hot and temperate environments. Under hot conditions fat is helpful as a source of readily available energy, to help with the increased energy demand for increased rate of respiration – panting. Under more temperate conditions, carbohydrates are the preferred source of energy as they are not as readily metabolised and do generate additional heat during metabolism (heat increment). Fat composition and quality is important for the breeder and progeny. Essential free fatty acids are required for cell membrane integrity, immune competence, fertility and embryonic development.

It is well documented that maternal feed fat composition directly influences the fat composition of the egg. This includes fatty acid profile and fat soluble vitamins. Wang et al. (2002) were able to show that maternal dietary polyunsaturated fatty acids (PUFAs) affected the PUFA levels in the spleen of progeny. They also demonstrated feeding Omega 3 fatty acids to the breeder resulted in higher Omega 3 fatty acid levels and immune responses in progeny. The immune response was measured by reduced wing swelling in birds injected with bovine serum albumin. Some other effects of vitamin supplementation will be addressed under the vitamin section.

Crude Protein and Amino Acids

Crude protein is a measurement of the nitrogen content of a feed or an ingredient, assuming that amino acids are the source of measured nitrogen. Crude protein does not give us any information about the amino acid composition and or availability. Modern nutritionists should only formulate feeds on digestible amino acid levels, using crude protein solely as means of easily analysing blended feeds to ensure correct blending.

Commercial feeds are generally still formulated to minimum crude protein levels, which results in feeds that, with the exception of TSAA (total sulphur amino acids), have amino acids significantly higher than required. The critical amino acids in breeder nutrition are methionine (TSAA), tryptophan, lysine and isoleucine. Methionine and tryptophan directly impact egg size and egg production. A corn/soy diet has excess lysine and isoleucine. When formulating to minimum crude protein levels, the lysine levels are often up to 40 per cent above requirement. Work by Lopez and Leeson (1995) clearly illustrated the negative effect of excess crude protein on fertility and DeBeer and Coon (2006) identified lysine and isoleucine as two amino acids directly affecting the fertility of breeders.

Minerals

All nutritionists are aware of the critical importance of the macro minerals; calcium, available phosphorus, etc. on bone and egg shell integrity. However, there is very good evidence that maternal levels of trace minerals especially zinc, manganese, copper and selenium impact levels in the egg and influence progeny. This is not new research; hen feeds deficient in zinc were shown by Edwards et al, (1959) to cause slow growth of chicks; Turk et al. (1959) to cause weak chicks, poor feathering and high mortality; and Kidd et al. (1992) showed supplementing with inorganic and/ or organic Zn increased levels of Zn in the bones and increased bone weight.

In addition to Zn's role in DNA and enzymes, Zn is particularly important in the young bird in the synthesis of two key proteins: collagen and keratin (Underwood and Suttle, 2001). Keratin is a structural protein in skin and feathers. Collagen is the major structural protein of internal tissues, including cartilage and bone. The cost of ensuring a sufficient and an available Zn source in breeder and early broiler feeds is minimal considering the impact of poor skeletal development and compromised immunity on the profitability of broiler production.

As with Zn, Mn, Cu and Se levels can be affected in the egg by maternal supplementation. Manganese is vital in embryonic and post-natal bone development. Cu is essential for reproduction and development. Se has a sparing effect on vitamin E as an antioxidant.

Vitamins

Although severe vitamin deficiencies will cause a wide variety of deformities and severely affect hatch, grossly underfeeding vitamins is not commonly seen in practice. It is the marginal deficiencies caused either by low supplementation, sources of questionable quality/availability and less dominant breeders consuming less than calculated feed quantities. The progeny will not exhibit classical deficiency syndromes, but they will not perform to their potential. Aviagen conducted a study to asses the impact on progeny where Vitamins E, K and B vitamins were supplemented at 20 per cent above Aviagen breeder recommendation. Although broiler body weights were only 20 grams heavier at term, mortality of the supplemented group was 2.2 per cent lower than the control birds, with a yield advantage of 0.2 per cent at 2kg body weight.

Egg yolk vitamin E levels were measured and a 50 per cent increase in α-tocopherol seen. In addition, candling clears at 60 and 64 weeks of age were 12.2 vs. 17.3 and 17.9 vs. 26.9 per cent per cent for the higher supplementation group. Adequate vitamin and trace mineral supplementation with quality vitamins and available minerals is an inexpensive way to ensure that the young chick is prepared for optimal skeletal growth and a healthy immune system, to help deal with challenges during brooding.

Amino Acids and Enteritis in Broiler Chickens

An intact healthy gut is vital to the health and welfare of young chickens. The gut as a multi-cellular organ; has absorptive enterocytes, goblet cells for mucin production; immune cells and an intrinsic neural system. The gut is the first line of defence against, dietary toxins and enteric pathogen invasion. According to Reed (2001) more than 65 per cent of dietary threonine is used in gut function; the threonine content of mucin is excess of 30 per cent. Enteritis results in considerable loss of mucin, studies have shown that threonine will be partitioned away from protein deposition to support the gut and mucin production.

To demonstrate the importance of threonine under enteric stress, a threonine dose response study was conducted to evaluate the impact of L-threonine on enteritis in broiler chickens. Birds were reared to 14 days on a coccidiostat-free, corn/soy feed with increasing levels of added L-threonine (at lysine to threonine ratios from 64 to 72). At 14 days, the birds were challenged with *Clostridium perfringens* and grown to 42 days of age. Body weights of the challenged birds at the higher threonine levels showed a positive response at 31 days and were similar at 42 days. Breast meat yield on the high treatment was similar between the challenged and control groups at 42 days.

This study supports the theory that animals will partition threonine away from muscle protein to support gut health and mucin production (Ajinomoto Heartland 2003).

Breeder nutrition directly affects the health, egg production and fertility of the birds. There is considerable evidence that breeder nutrition affects the progeny. Although we seldom see the gross deformities of severe nutrient deficiencies, it would appear that current nutrient levels, particularly trace mineral and certain vitamins are marginal particularly for those less aggressive birds that are unable to consume their expected feed allotment. It is these subtle deficiencies that could benefit from proper supplementation of these nutrients.

Overfeeding of crude protein:

1. Costs money; a recent evaluation indicated around $3.00 per ton.
2. Results in additional heat increment, which contributes to the heat load under heat stress conditions. New high yield breeds are particular sensitive to heat stress and nutritionists need to help minimise this stress.
3. Negatively affects fertility
4. Excess protein is excreted, and elevates the nitrogen in the faeces. High nitrogen in the litter results in foot pad lesions and excess ammonia, which can irritate mucus membranes of the eyes a respiratory tract.

Breeder hens are producing the broiler chicks that are the basis of our business. Feeding them correctly will result in healthy, virile chicks, which will have the immune system and skeletal framework required to deal with challenges and achieve their growth potential. The increase in cost of feeding a breeder hen properly compared to the total cost of producing a mature broiler chicken is marginal.

Saving cents may be costing you Rands.

Energy Costs Associated with Commercial Broiler Production

Data recently compiled at the Applied Broiler Research Unit may be of value in assessing your farms' energy demand and (based on your costs for fuel and electricity) monetary expense to meet this demand.

Housing and Management Practices

The information presented represents data from 38 consecutive flocks of straight run broiler chickens grown at the Applied Broiler Research Unit during the period October 1996 through June 2003. All flocks were grown for the same integrator under a standard broiler production contract.

The houses were all 40 × 400 ft. Two houses (1 and 3) featured conventional cross-ventilation with low-pressure foggers, while the other two houses were curtain-sided and tunnel ventilated. One tunnel ventilated house (4) had evaporative cooling pads and the other (2) had an experimental sprinkler system. Detailed descriptions of the houses, environmental control systems, sprinkler system, and housing modifications was given by Berry et al. (1991), Xin et al. (1993) and Tabler and Berry (2001). Management practices were the same in all houses and the farm manger was the same individual throughout the study period. Half of the 38 flocks were grown for 49 days or less while the other half were grown for more than 49 days. The youngest flock was 39 days at harvest while the oldest was harvested at 57 days.

Propane Usage

The lower fuel consumption in House 3 during the winters of 1998 and 1999 was due to use of an experimental wood-burning pellet furnace.

House 4 was the most challenging house to control from a management standpoint since it had more ammonia than any other house. This increased ammonia required increased ventilation to maintain the proper environment resulting in increased gas consumption during cooler periods of the year. Although House 4 consumed the most fuel during the seven-year period, it should be noted that when the 1998 and 1999 data were ignored, tunnel ventilated houses consumed only 2% more fuel than did conventional houses. Also, if the ammonia problem in house 4 could be solved, tunnel ventilated houses would likely consume less fuel than conventional houses.

Electricity Usage

The summer of 2000 was the most costly in terms of electricity usage followed by the summer of 2001. However, unlike propane usage, each house accounted for an equal amount of electricity usage (25%) during the 7-year period. The increased electrical demand in House 3 during the winters of 1998 and 1999 are again associated with use of the wood-burning pellet furnace in that house.

Even though House 1 showed the highest kWh usage as compared to the other houses, there was less than 185 kWh difference between houses and the houses were the same when compared on a percentage-of-use basis. Electricity usage for the farm over the 7-yr period averaged 12,617 kWh per flock. Based on 5.5 flocks per year the farm would have used 69,394 kWh per year or 17,348 kWh per house per year. If electricity costs $0.06 cents per kWh, electricity costs would come to $4,164 for the farm or $1041 per house.

Total Energy Costs

However, because fuel costs are much greater than electricity, growers have a much more serious problem dealing with high fuel bills in the winter than they do the electric bill in the summer.

Contract growers face numerous challenges associated with raising broilers. One significant challenge is the monetary expense related to fuel and electricity costs. Energy data from 38 consecutive flocks of straight run broilers over a seven-year period at the Applied Broiler Energy Unit indicate that approximately 25% of the gross farm income is required to pay the annual propane and electricity bills and that propane costs may be roughly four to five times the cost of electricity. While energy costs will vary somewhat from farm-to-farm, the wise use of energy should be a priority for all growers.

Nutrient Digestibility of Broiler Feeds Containing Different Levels of Variously Processed Rice Bran Stored for Different Periods

Nutrient digestibility of broiler feeds containing different levels of variously processed rice bran stored for varying periods was determined. A total of 444 Hubbard male chicks were used to conduct four trials.

Each trial was carried out on 111 chicks to determine digestibility of 36 different feeds. Chicks of 5 wk age were fed feeds containing raw, roasted, and extruded rice bran treated with antioxidant, Bianox Dry

(0, 125, 250 g/ton), stored for a periods of 0, 4, 8, and 12 mo and used at levels of 0, 10, 20, and 30% in feeds.

Digestibility coefficients for fat and fibre of feeds were determined. Increasing storage periods of rice bran significantly reduced the fat digestibility of feed, whereas no difference in fibre digestibility was observed. Processing of rice bran by extrusion cooking significantly increased digestibility of fat even used at higher levels in broiler feeds. Interaction of storage, processing, and levels was significant for fat digestibility.

Treatments of rice bran by different levels of antioxidant had no effect on digestibility of fat and fibre when incorporated in broiler feed.

Meeting the Nutrient Requirements of Broiler Breeders

Major changes have taken place with the feeding of broiler breeders in recent years. One of the most significant of these changes was separate sex feeding. With such an approach to feeding it is now possible to more closely control the weight of the pullet coming into production and to be confident that, with challenge feeding, the additional feed is being used to maximise egg output rather than put extra weight on already overweight cockerels. By being able to precisely control cockerel weight, marked improvements in hatch of total eggs set, with reduced cockerel numbers, is being attained.

With separate sex feeding, changes in ration formulation to more precisely meet nutrient requirements of males and females are now possible. Also more attention can now be paid to flock uniformity and by switching to everyday feeding, uniform pullet flocks that reach peak production in 4 to 5 weeks, and produce 180 to 190 eggs per breeder, is not uncommon.

While there are many points to consider with respect to management, feeding, health care, photoperiod, etc., that can have a bearing on the economics of production, the present article will deal mainly with feeding to more precisely meet the energy and protein requirements of the modern day broiler breeder.

Body Weight and Condition

While body weight recommendations are readily available for the various commercial breeders on the market today, often there is little information given on how to keep problem flocks close to these suggested weights. A 6 week old commercial female broiler will weigh around 1800g while a weight of 770g is the target weight for a broiler breeder pullet of the same age. Hence, it is evident that overweight, not underweight pullets are usually the problem during the growing period. Pullets that are overweight during rearing should not have their feed allotment reduced in order to bring their weight in line with suggested target weights. Rather, feed should be held at the same level until body weight comes in line with that recommended for a given age. Then adjustments in feed allowance should be made to hold pullets to the desired weight curve.

Another important point is to ensure that pullets, as they approach mature pullet weight, are carrying sufficient body flesh. This can be observed by checking fleshing over the keel. Pullets that are lacking in flesh are immature birds and will be delayed coming into production. Increase feed, if necessary, to ensure well fleshed pullets and delay light stimulation of the flock until proper body condition has been achieved, even though body weight may be slightly in excess of the weight guide.

Nutrient Requirements

Two of the major and most costly nutrients required by the breeder are energy and protein. While protein requirements have often been the major consideration of most producers, usually it is energy, the most costly dietary nutrient, which has the most influence on performance.

Energy

Energy is often referred to as "The Fire of Life". Feed supplies energy, which in turn is required for growth, egg production, maintenance, vital life functions and activity. In developing feeding programmes for broiler breeders consideration should be given to body size, as this is a major factor, along with environmental temperature, influencing the maintenance energy requirement. Production of a product is the other major consideration. This may include a growth component but the main consideration is egg mass output.

Maintenance

The maintenance energy requirement is affected by body size, environmental temperature and level of activity. In order to calculate energy requirements a knowledge of the calorie requirement per unit body size (surface area), growth rate, and egg mass output must be known. Environmental temperature must also be taken into account as it will affect the energy needs for maintenance.

This is important for as will be discussed, the energy requirement for maintenance makes up by far the greatest percentage of the birds total energy requirement. A number of equations have been developed which give reasonable estimates of the total energy requirement of a broiler breeder. However, as with any predicted estimate, such values must be evaluated for each and every situation.

Energy Requirements for Pullets

It is important that growers recognise that the major factor influencing feed intake is the bird's need to meet it's maintenance energy requirement. Since environmental temperature is the main factor influencing maintenance energy requirement (other than body weight), pen temperature should be closely monitored and feed intake adjusted accordingly if significant temperature changes result. While precise changes in feed allotment, to account for changes in environmental temperature, are difficult to predict accurately, a reasonable estimate would be that for every 1°C change in pen temperature through a temperature range of 15 to 30°C, the bird will alter its feed intake by 3g/d. However at high or low temperatures the change in feed intake would be significantly greater.

Energy Requirement for Broiler Breeders

This decreases noticeably at peak production and continues to peak egg mass when the hen would be partitioning a significant proportion of its' energy intake to meet its requirement for egg mass output.

Again if one is to precisely meet the breeder's energy requirement, factors influencing this requirement must be considered. One is the increase in body weight and this is apparent by the steady increase in maintenance energy requirement as the production cycle progresses. However, total energy requirement peaks around peak egg mass production and then declines as the requirement for egg mass production is reduced as egg production decreases.

It is important that the producer recognise the fact that only around 20% of the energy intake of the hen is diverted into egg production and increased body weight (which would be fairly small). Since the hen will preferentially partition nutrients to meet her requirement for maintenance, if her feed allowance is not sufficient to meet her total energy requirement, egg mass output and thus egg production and/or egg size will be reduced.

Influence of Feather Condition or Energy Requirements

While pen temperature has a significant effect on maintenance energy requirement, the feather condition of the hen can significantly influence the affect of environmental temperature. A marked increase in feed intake is noted with birds as they lose feathers, which often occurs with aging as well as poorly managed flocks. Allowance must be made for feather condition when estimating the calorie needs of a flock if maximum production and egg size are to be maintained.

Reducing Feed Intake Past Peak Production

It is common practice to reduce feed allowance to a flock shortly after peak production is attained. This may be responsible for some of the dips in production noted shortly after peak is attained.

Another error that a number of producers make is to reduce feed allowance in relation to the drop in egg production, (eg.) if production drops 5%, feed allowance is reduced by a similar amount.

An average egg of around 65g requires approximately 140kcal of metabolised energy for its production. If feed allowance is 160g/b/d at peak production, with a diet containing 2800 kcal/kg, this will result in a daily intake of 448 kcal ME/b/d. If 140 kcal are required for the production of one egg this means that approximately 31% of the hen's energy intake is going towards egg production at this time. Egg mass output at 85% production and with a 65g egg = 55.3 g/h/d. If production drops to 80%, and during this time egg weights increases by 1g, egg mass output would be 52.8g/h/d. Thus egg mass output would drop;

$$\frac{55.3 \div 52.8}{55.3} \times 100 = 4.5\%$$

Hence, the energy required per hen per day for egg production should drop;

$$\frac{4.5 \times 140}{100} = 6.3\text{kcal}$$

Thus for a 4.5% drop in egg mass output total dietary energy requirement has only dropped

$$\frac{6.3}{448} \times 100 = 1.4\%$$

...(not the 5% that one might estimate from the drop in egg production). It should be obvious that decreased feed intake after peak production has to be precisely calculated as the hen's requirement is not related directly to the drop in egg mass output. From the above discussion it should be apparent that there is a strong possibility that the broiler breeder is deficient in energy intake up to and perhaps beyond peak egg mass production. As already mentioned any deficiency in energy intake will readily translate into reduced egg numbers or smaller eggs or both.

Protein: Since the maintenance requirement for energy is so high, in relation to the breeders total energy requirement, energy intake is one of the main factors influencing the reproductive performance of broiler breeders. Thus changes are often made in daily feed allotment in order to try and maintain an optimum intake of energy.

However, little attention is paid to the amount of protein consumed by the hen. Most breeder diets contain between 16 and 18% protein. With variable feed intakes during the laying period, daily protein intake obviously varies. Many breeders are fed up to 770g of feed per day (or more). Thus protein intake could range from 25.6 to 28.8g per day, and in many cases higher).

Protein Required for Egg Production

What is the calculated protein requirement for the broiler breeder? Using values generated for commercial layers, the protein required to produce a 65g egg (containing 770g of protein) should be around 7.8÷.55 (suggested efficiency of dietary protein utilisation for egg production) = 14.2g.

Protein Required for Maintenance

With an estimated total endogenous loss of nitrogen, including feather loss, to be 280 mg/kg of body weight to the 0.75 power, a 3.5kg hen would require 3.5^{75} X 280 = 717mg of nitrogen per day to meet her maintenance requirement. Converting this to protein required would give (.717 X 6.25) or 4.48g of protein. Assuming the hen is 55% efficient in converting dietary protein to body proteins the hen would require 4.48 ÷ .55 = 8.15g of protein intake per day to meet her maintenance protein requirement.

Total Daily Protein Requirement of a Broiler Breeder

Considering egg production and maintenance the protein requirement would be 14.2 + 8.15 = 22.4g/h/d. This should be sufficient for a hen to lay a 65g egg every day. While no allowance has been made for body weight gain, this should be minimal after peak production and besides much of this gain will be fat deposition and thus a minimum of body protein would be deposited. Since every bird in a flock is not laying an egg a day the protein required for egg production per day is only the average percent production times the estimated 14.2g of protein required for the 65g egg. Thus the daily protein required of a flock of broiler breeders would be significantly less than 22.4 g/b/d. While the above values are only estimates of protein requirements, they are reasonably good estimates based on values generated mainly from egg production hen values.

Partitioning the Breeders Protein Requirement

While it was shown that around 80% of the energy consumed is partitioned to meeting the hens requirement for maintenance (including weight gain and activity) it can be estimated from the values generated for protein requirements that, (8.15 ÷ 14.2 X 100) approximately 57% of the protein intake of the breeder is going to meet its' requirement for egg mass production. Thus the main factor influencing the protein requirement of the broiler breeder is egg mass output, not body maintenance as with energy.

Estimated Protein Requirements for Flock Breeders

Percent Dietary Protein to Meet Requirements: This would translate into dietary protein levels of from 7 to 11.4% to meet these estimated requirements, assuming an average feed intake of 160g/b/d. While a lot of assumptions and estimates have been made in generating the above values the low levels of dietary protein suggested are not too far removed from the estimates suggested by Bowmaker and Gous (1989), Harms and Ivey (1992) and Lopez and Leeson 1993).

As can be noted in 10% dietary protein resulted in comparable egg production from the report of Lopez and Leeson (1993) as compared to a 16% protein control diet. Indeed by the end of the trial, egg numbers produced were similar for both these treatments. However, based on the body weight curves 10% dietary protein was not high enough to maintain body weight at acceptable levels. As has been reported on numerous occasions with egg production type hens low dietary protein levels, while supporting good egg production usually result in smaller egg weights regardless of essential amino acid supplementation.

Lopez and Leeson (1993) also demonstrated and increase in hatchability with lower protein diets. It is interesting to note the number of reports that have shown the detrimental effects of high protein intakes for broiler breeders, yet many in the industry still insist on feeding high protein breeder diets.Pearson and Herron (1981) reported that a ratio of dietary protein to energy of higher than 15g:1M joules resulted in reduced hatchability of breeders. This is of interest as converting their ratio to protein per 100 calories gives a value of 6.28 which is in line with the ratio of protein to energy reported by Spratt and Leeson (1987) to maximise chick hatch weight.

There appears to be a lot of evidence to suggest that most broiler breeders are being subjected to excessive intakes of dietary protein. Not only is such a practice detrimental to performance, but it is uneconomical as well as resulting in a greater potential pollution problem with higher levels of nitrogen in the litter from such flocks.

The argument is still heard that "I have to continually increase daily feed allowance in order to maximise egg production", and/or "I had to increase dietary protein levels to increase egg number and size". As has been clearly demonstrated the breeder's main nutrient requirement is for energy and around 80% of this (under good management conditions) goes just to maintain the bird. If the hen's intake of feed is not sufficient to meet her maintenance energy requirement, dietary protein will be preferentially burned for energy purposes. Hence, a level of dietary protein which should have readily met the hen's total requirement is now deficient with respect to meeting the breeder's requirement for egg mass production. If dietary protein is increased, with such a situation, often an increase in egg mass output is noted. Also if an increase in feed allowance is made a response in performance is usually also noted as more energy is available to meet energy requirements, thus leaving a greater amount of dietary protein to be utilised for egg mass output

What often happens is that dietary protein levels are increased to the point where protein is in excess of requirements and thus nitrogen excretion is increased. It requires a substantial amount of energy to synthesise and excrete uric acid, the nitrogen excretory product of birds. Thus, as dietary protein levels are increased, protein intake increases, uric acid excretion increases and the energy requirement of the bird significantly increases.

With such a situation production responses are often noted with excessive allowances of feed. If a flock is not attaining expected egg numbers and size with a daily feed allowance of 150 to 160 g/b/d, one should take a critical look at possible management factors before changing diet composition or significantly increasing feed allowance. However, it must be remembered that some well managed uniform flocks are reaching production peaks of over 85% and maintaining good production longer than were flocks a few years ago.

Such flocks may require more than the normally recommended level of feed allowance in order to meet their energy requirements. While it is true that as egg mass output increases and a larger percentage of protein intake is partitioned into egg production, one might question whether dietary protein level should be increased. However, with the increased intake of protein as feed allowance is increased, intake of dietary protein is seldom, if ever a problem with broiler breeders.

As mentioned previously there are many good production models available for estimating energy and protein requirements of breeders and in many cases these are considered by nutritionists. The intent of the present article was not to try and compete with the nutritionist regarding diet formulation, but rather to try and point out to producers avenues to pursue and reasons why they may not be achieving optimum performance from their flocks.

It is often stated that seldom is the diet at fault but rather it is the feeding programme or the management conditions under which the diet is being fed which is the problem. This is especially true for broiler breeders where the nutrient intake and requirements of the birds is very much under the control of the flock manger.

Broiler breeders are fed to maximise the production of saleable chicks per bird. Considerations of some of the points raised in the present article might help some producers to increase this number.

The Only Good Broiler Breeder Egg is a Fertilized Egg

Hatchability is around eight percentage points lower than fertility because many chick embryos are usually lost during incubation. For example, even if 93 percent of the eggs laid are fertilized, then under normal incubation conditions only 85 percent of the eggs will hatch. This example illustrates how fertility must be very good to get above average hatchability and hatch bonus pay.

Breeders need to be kept under ideal conditions for maximum life of flock fertility. The chicken's reproductive system is very sensitive to the bird's environment, and under poor conditions the reproductive system will dwindle. For example, the environment can cause a rooster's testes to increase or decrease in size by several hundred fold. But, before we can understand which management factors influence fertility, we must first examine the fascinating process of fertilization in poultry.

Fertilization in any animal depends on production of eggs from the female and sperm from the male. A problem with either sperm or egg production can decrease fertility. The rooster's reproductive system is simple when compared to humans or other mammals. The rooster does not have a prostate gland or any of the accessory reproductive glands. Like all other animals, chicken sperm carry the genetic material from the rooster and are produced within the testes. The rooster has two very large testicles within the abdominal cavity on each side of the backbone. After sperm leave the testes, they enter the epididymis, where they gain the ability to swim. Next, the sperm enter the vas deferens, where they are stored until the rooster mates with a hen.

Sperm formation takes about 15 days. The rooster's semen contains around 5 billion sperm per cc, about 40 times as much as that of a human. Once a rooster is mature and if he is maintained properly, he will manufacture about 35,000 sperm every second of his life. However, just like the males of many animal species, the fertilizing potential of roosters varies, even within a flock. For example, some roosters are extremely fertile and create a maximum number of quality sperm; other roosters are subfertile and do not make enough good sperm. This variation in rooster quality is caused by management, environment, nutrition, and genetics.

The hen does not produce nearly as many eggs as the rooster produces sperm, but during her 40 weeks of production, the broiler breeder hen lays about 180 eggs. Egg formation requires about 25 hours. Since egg formation requires more than 24 hours, even the best hens cannot lay an egg every day in succession throughout their productive life. As is the case with roosters, some hens are more productive than others, and management has a major impact on variability among hens.

The hen's reproductive system can be divided into two major components: the ovary and the oviduct. The ovary produces the egg yolk. The oviduct adds the white, shell membranes, and shell to the yolk to complete egg formation.

The hen has only one ovary, which is on the left side of her abdomen. The ovary has several thousand ova (egg yolks) in different stages of development and looks like a bunch of grapes. Very immature yolks contain only genetic material from the hen, and as the yolks grow to around 1 mm in diameter, they become white. If the hen is managed properly, many of these developing egg yolks will mature in about 19 days into large, 35 mm, yellow yolks. As the egg yolk develops it will get water, sugars, fats, proteins, vitamins, and minerals from the hen's blood. These are all necessary for the embryo to develop. The egg yolk is surrounded by the perivitelline membrane. This keeps all of these nutrients in a ball-shaped package. One particularly visible region of the perivitelline membrane is the germinal disc. The germinal disc is a small white dot about half the size of a pencil eraser on the surface of the yellow egg yolk. Fertilization takes place here, and embryonic development begins.

When the egg yolk is mature, it leaves the ovary, and within 20 minutes it is captured by the infundibulum, the first part of the oviduct. Here fertilization takes place. Following mating, sperm enter the hen's oviduct and are stored within sperm storage glands. Only sperm that can swim will enter these sperm storage sites. These glands can store more than half a million sperm. Sperm can remain alive in these glands and fertilize eggs for up to 3 weeks. A hen will have maximum fertility for only about 3 to 4 days after one mating. For this reason, the male-to-female ratio in a flock must be enough to ensure mating of every hen every 3 days or so. Sperm do not break through the eggshell. Instead they travel up the oviduct to the infundibulum to join with the egg yolk.

The sperm bind to the perivitelline membrane and make a hole as they enter the egg. Hundreds of sperm may enter the yolk. As a matter of fact, the more sperm that enter the yolk, the more likely the egg will be fertilized. Around 30 sperm must enter the egg near the germinal disc to insure a 95 percent chance of fertilization. While it is true that only one sperm is necessary to fertilize an egg, the probability of an egg's being fertilized by only one sperm's reaching and penetrating it is very low.

After about 15 minutes, the yolk leaves the infundibulum (fertilized or not) and receives the egg white, shell membranes, and shell over the next several hours from the magnum, isthmus, and uterus sections of the oviduct. When the hen lays a fertilized egg, the chick embryo has already developed for about 25 hours into approximately 20,000 embryonic cells and is a live, breathing organism. If this fertilized egg is handled properly before and during incubation, a healthy baby chick is the result.

This publication is a joint effort of the Mississippi State University Extension Service and the Mississippi Agricultural and Forestry Experiment Station.

Weighing Broiler Breeder Females Post-Feeding

Obtaining accurate body weights is a critical part of the process of rearing replacement broiler breeder pullets and managing breeder hens and males. From the first few weeks of age in the pullet house, all feed allocations are determined by the bird's weekly weight gains.

Obtaining accurate body weights is very important to maintaining uniformity, body conformation and the overall development of pullets and young cockerels. Research has shown that accurately and uniformly controlling body weight of both replacement breeders and breeders in the hen house will result in improved performance parameters.

In the United States, the majority of poultry integrators rear pullets on some version of a skip-a-day feed programme in order to control body weight among all the birds in a house. Under our current housing conditions, skip-a-day feed programmes are the best way to uniformly distribute feed to all birds simultaneously in an effort to maintain body weight uniformity. However, the presence of feed in the crop or digestive tract will inflate the actual body weight of the birds and skew feed allotments.

Therefore, replacement breeders are typically weighed on off-feed days to normalise the data and not confound body weights with either the presence or absence of feed in the crop or digestive tract. This allows for body weight measurements to be consistent from week to week without regard for feed clean up time and the presence or absence of feed in the crop. Therefore, each week pullets and cockerels are weighed with an empty crop and digestive tract. This process continues until birds are moved to the hen house and feeding begins on an everyday basis. These weights are considered to be 'empty' weights.

In the hen house, most commercial producers move from a skip-a-day to an everyday feed programme as hens are brought into production. Feed is often provided daily in the early morning hours shortly after the lights are turned on. While feeding hens everyday in the hen house has proven to be an effective management tool, birds cannot be weighed on 'off feed' days.

This has led to the concern over whether hen weights are truly reflective of the actually body weight and mass. Consequently, current industry recommendations are designed to address this issue and suggest producers weigh breeders late in the afternoon hours to obtain the 'empty' weights. This allows any feed consumed to have time to pass through the birds digestive system and therefore create an 'empty' weight situation for weighing purposes. In breeders this can be further complicated by the fact that the majority of egg production occurs in the morning hours following feed cleanup which would result in additionally body weight loss.

To address this issue, a research project was designed to weigh breeders at various intervals during the day to determine the best time to weigh birds to most accurately reflect actual body weight gains.

When to Weigh Breeders

Birds used in this study were housed at the University of Arkansas Broiler Breeder Research Farm. A single pen of breeders containing 71 hens was used for this study and during each weigh period, all hens were corralled in a catch pen with each hen weighed individually so that no sampling error could affect the results. All hens were weighed prior to daily feeding and again at feed clean-up time. Additional bird weights were obtained at 2, 4, 6, 8 and 10 hours following feed clean-up. This process took place on the same birds at 24, 28, 34 and 41 weeks of age. These age periods represented pre-laying, pre-peak, peak and post-peak in production stages of life. It was previously believed that hens would lose body weight throughout the day to approach the 'empty' weights found prior to feeding. However, these data make it apparent that the passing of feed and the consumption of water appear to offset each other and allow the hen to maintain a near constant body weight through 10 hours following feed clean-up. Body weights obtained prior to feeding would be the only weights that could be considered 'empty' weights as they were obtained immediately after lights came on in the morning and are a reflection of body weight loss due to feed and water passage occurring during the dark hours.

These results would allow breeder service technicians to weigh breeders in the hen house at any time following feed cleanup and that the data would be consistent with body weights obtained at any time throughout the day. These data will allow technicians to be more productive in a given day in regards to scheduling weighing of breeders in the hen house.

How to Weigh Birds

When weighing birds, it is often recommended to weigh all birds caught in a catch pen and not weigh a specific number of birds to meet a given criterion. This has been the recommendation for broilers in research trials but has not been evaluated in replacement pullets and breeders. As part of this project, body weights were recorded for each hen in the order they were caught in the catch pen. For each age group and for each time interval previously mentioned, this resulted in 40 incidences of weighing all birds in a catch pen.

For instance, if 60 birds are caught in a catch pen and only the first 50 are weighed because that meets the minimum number needed, then the body weight recorded would not be reflective of the actual weight of the birds caught or the birds in the flock. If this occurs with pullets and feed allotments are determined based upon these body weights, then inaccurate feed allotments could be provided and less control over flock body weight would be the result.

Summary

1. When weighing broiler breeders in the hen house, accurate and consistent body weights can be achieved by weighing birds at any time after feed clean-up. There is no advantage to waiting for feed passage in an attempt to obtain 'empty' weights in breeders during the afternoon hours.
2. When weighing birds caught in catch pens, it is important to weigh all birds caught in the pen and not stop at a predetermined number

of birds. The last birds caught will be the smallest birds and need to be included in the final group weight to most accurately determine the average body weight of the birds in a flock.

Health Implications for Higher Density Broiler Production

Such actions are making us acutely aware of the "law of diminishing returns." We are now faced with a more accurate evaluation of the application of the "law" to the broiler chicken of today, the facility in which the chicken is grown, the feed the chicken is fed and the grower who manages it all.

The broiler chicken of today is truly remarkable. The superior genetics has resulted in increased body weight gains, increased yields, and increased efficiency of feed utilisation. In Western Canada, managing the rapid growth rate has become our number one priority. To accomplish this task, the grow facilities are usually equipped with computers to consistently control the bird's environment. Feed formulations have been improved. Water and feed delivery systems are being improved. All of this investment in equipment and time will ensure the bird reaches its full potential. But, will this still be the case with increased bird densities?

The health implications for higher density broiler production are significant and must be taken into account. Bird welfare, in my opinion, must also be considered. Bird welfare, is not going to be, it is, extremely important in terms of meeting the consumer demands of today and tomorrow. What happens when we add more kilograms of live weight to a given, fixed amount of floor space? As we examine the possibilities, we should always evaluate our FLAWS.

With increased density, feed and water will become more difficult for each bird to access. This will lead to reduced performance on each normal bird. Furthermore, birds which may only have a marginal disability will become less able to compete as the stocking density rises.

Poorer litter conditions associated with higher moisture content occur with higher stocking densities. Higher litter moisture will "increase the wave of insult" by increasing the concentration of pathogenic organisms. Reduced access to feed containing anticoccidials/ antibacterials together with increased exposure to oocysts/bacteria may result in clinical coccidiosis and/or bacterial infections. Poor litter conditions can also exacerbate leg disorders experienced by chickens.

Poorer litter conditions, reduced access to feed, and an increased demand for fresh air may result in an increase incidence of pulmonary/ cardiovascular disease. Presently, we are faced with an ever increasing incidence of pulmonary hypertension (ascites). By increasing the stocking density we will be increasing the demand for vital oxygen. Increasing the demand for oxygen will add more pressure to the bird's pulmonary and cardiovascular systems. Increasing stocking density will increase the likelihood of a bird being scratched. This coupled with an increase in litter moisture will set up the possibility for an increase in the incidence of type II cellulitis. There are many "roads that lead to the cities of ascites and cellulitis" and increasing the stocking densities will only make "these roads easier to travel."

The ability to vaccinate our birds via the drinking water will be compromised by increasing the stocking density. Poorly vaccinated flocks are more prone to vaccine "rolling" reactions and are more prone to disease. Increased stocking densities will increase stress. Increased stress will manifest itself in many ways, most commonly as a reduction in overall performance. Increased stress will result in an increased susceptibility to the common broiler diseases in a given geographical area and may open the door for new and re-emerging diseases. If we increase the stocking density of our broiler chickens beyond our capabilities as broiler farm managers, then we will suffer losses. If we increase our capabilities as broiler farm mangers, then we will be able to grow more chickens in less space. Let us first increase our capabilities, then add more chickens.

As a poultry veterinarian today, the majority of my work has revolved around solving "problems" that involve more than one etiology or cause. It clearly involves concentrating on prevention. "Magic bullets" do not exist. The diseases that we are faced with "preventing" have multiple and convoluted causes and involve the understanding of all disciplines of broiler production. Our broiler chickens have served us extremely well in the past and by all present indications our broiler chickens today are performing even better. More commitment will be required by the entire broiler production team. This team will be required to operate and work together in an orchestral fashion in order to achieve success.

Chapter 4

Dietary Protein Formulation

Necrotic enteritis (NE) is a widespread and economically important bacterial disease in modern broiler flocks (Van Der Sluis, 2000). The subclinical form of the disease is more common than clinical outbreaks in broiler flocks (Kaldhusdal, 2000). The condition is not usually detected due to the absence of clear clinical signs; therefore, it is not treated and prevails unnoticed apart from a poor growth performance (Lovland and Kaldhusdal, 2001), wet litter conditions (Williams, 2005), and the possible contamination of poultry products for human consumption (Craven et al., 2001). The financial cost of NE has been estimated to be US $2.6 billion per year to the world's poultry industry (Van Der Sluis, 2000).

In subclinical NE, the major pathological changes occur in the small intestine (Gholamiandehkordi et al., 2007) and the liver. Intestinal *Clostridium perfringens* counts (Kaldhusdal and Hofshagen, 1992) and intestinal *C. perfringens* α-toxin levels are also increased. The causative organism, *C. perfringens*, is ubiquitous and found in soil, dust, feces, feed, used poultry litter, the intestines of most healthy animals, and humans. *Clostridium perfringens* also is a commensal bacterium of chicken intestines (Ewing and Cole, 1994) and so there are further predisposing factors that alter the intestinal balance in favour of the proliferation of the causative bacteria and allow them to migrate to the upper intestines. Apajalahti et al. (2001) identified the diet as the strongest determinant of the cecal bacterial community, so the diet composition may influence the susceptibility of broiler chickens to subclinical NE, although there is a lack of direct experimental evidence.

However, there is evidence that different dietary protein sources affect the proliferation of *C. perfringens* within the cecum (Drew et al.,

2004) and in the ileum when birds are orally dosed with these bacteria. Also, studies have indicated that the dietary amino acid balance may influence the proliferation of *C. perfringens*.

Most of the published experimental information on NE has reproduced the disease by dosing the birds with pathogenic strains of *C. perfringens* isolated from clinically diseased birds. Although the various protocols differ, most of these experiments involved keeping a small number of birds in cages with frequent dosing of the pathogen. The majority of the published work that has examined dietary factors that predispose birds to NE has also used much higher inclusion rates of the individual feedstuffs than would be used in proprietary feeds. All of these conditions are different from practical broiler chicken rearing methods.

However, Lovland et al. (2003) reported that a mild or subclinical NE initiates spontaneously without dosing the birds with pathogenic *C. perfringens*. This innate infection usually occurs when the appropriate predisposing factors (unmedicated diets, putative predisposing feeding regimens, housing birds on litter) are provided. This method of reproducing subclinical NE therefore provides the possibility of studying the effects of dietary factors under conditions that are directly related to commercial production methods.

There is a need to determine whether dietary protein supply affects the incidence of subclinical NE in practical broiler growth conditions. Dietary protein supply is a major variable in the formulation of poultry feeds and local and world price variations can result in large changes in the sources and quality of the protein used. Soy is the most frequently used protein concentrate in poultry feed formulations.

Experimental evidence suggests that high contents of dietary fish meal increase the intestinal *C. perfringens* populations in pathogen-dosed birds (Drew et al., 2004). Wilkie et al. (2005) identified that birds fed diets with high potato protein inclusion rates also had higher counts of *C. perfringens* when compared with birds fed other plant protein diets. However, the effect of potato protein and fish meal, in comparison to soybean meal, on naturally occurring spontaneous subclinical NE has not been investigated.

Therefore, the objective of this experiment was to identify the effects of 3 different dietary protein concentrates: potato protein concentrate (PPC), fish meal, or a mixture of soybean meal and full-fat soy in nutritionally complete diets, with similar protein contents (CP 21%), on the incidence of subclinical NE in male broiler chickens reared in comparatively large flocks and in conditions close to practical production methods.

Materials and Methods

Dietary Treatments and Experimental Design

Three practical broiler grower diets were formulated to be nutritionally complete for macronutrients (Aviagen, 2007a) and the vitamins and minerals were provided with a proprietary vitamin and mineral premix that either met or exceeded the NRC (1994) recommendations for broiler chickens between 15 to 31 d of age. Diets were corn-based, but a major proportion of the additional protein supply was provided by 1 of 3 protein concentrates: PPC (CP 79%) provided 58% of the total CP in the potato-based diet and fish meal (South American origin; CP 66%) provided 58% of CP in the fish-based diet. The soy protein-based diet was composed of a mix of soybean meal (dehulled; CP 48%) and toasted full-fat soy (CP 36%) and these 2 soy proteins together provided 61% of the total CP in the diet. All 3 diets were formulated to have similar contents of calculated ME (3.1 mcal/kg), CP (21%), Lys (1.3%), Met and Cys (0.94%), and Thr (1.02%). However the chemical analysis at the end of feed processing (mixing and pelleting) indicated that there were differences in amino acid, starch, lipid concentrations, and trypsin inhibitor activity between the treatment diets. No antibiotic growth promoters or anticoccidial drugs were used in the diets. The diets were all provided as 3-mm-diameter pellets. Treatment diets were compared using 6 replicate pens for each treatment. Eighteen pens were used within 2 adjacent environmentally controlled rooms. The dietary treatments were randomly allocated to pens within 6 positional blocks (3 per room).

Broiler Chicken Management and Feeding

The broiler chicken experiment was conducted under the guidance of the Research Ethics Committee of Harper Adams University College. A total of 1,300 one-day-old male Ross 308 broiler chickens were reared in a solid-floored pen as a single flock in an environmentally controlled house. Adequate feeders and drinkers were provided for the age and the number of birds. For the first 15 d, birds were fed a proprietary, nutritionally complete broiler starter feed formulation that was unmedicated (without in-feed antibiotics or coccidiostat). On 15 d post-hatch, the birds were weighed and 70 birds were randomly allocated to each of the 18 pens (altogether 1,260 birds were allocated to the experiment). The floor area of (1.5 m × 3 m) each pen was covered with wood shavings that comprised 4 parts new wood shavings to 1 part reused litter material from a previous poultry flock that did not have a history of clinical NE but some subclinical NE and subclinical coccidiosis would have been expected. The birds were fed the experimental diets for the following 16 d. Feed and water were provided ad libitum during the experimental period. Feed levels were kept low in the hanging tube feeders to minimise wastage. Light was provided for 23 h per day with controlled temperature and humidity. The growth and feed intakes of the birds were recorded over the experimental period. Mortality was recorded daily.

Data Collection

Lesion Scoring

On 27, 28, 29, and 30 d of age, 8 birds from each replicate pen (2 birds per day) were selected at random and killed by cervical dislocation and 5 mL of blood was collected in glass tubes from the jugular vein. The liver of each bird was examined for the presence or absence of lesions of hepatitis or cholangiohepatitis that were consistent with the pathological changes of NE (Lovland and Kaldhusdal, 1999; Sasaki et al., 2000). The intestinal tract was removed and the duodenal loop was separated and retained. Three further 8-cm sections were taken from the rest of the small intestine; the first of these sections was the proximal jejunum, the second section was either side of Meckel's diverticulum (mid small intestine), and the third section was the terminal ileum.

All 4 sections (including the duodenal loop) were immediately incised, washed in normal saline, and the mucosal surfaces were inspected and necrotic and hemorrhagic lesions were scored. Proximal jejunal sections were directly stored at "20°C. Sections (1 cm^2) from the jejunum that had focal necroses were fixed in 10% formal saline and after 2 d, these sections were rinsed with water and placed in 70% ethanol for histopathological diagnosis. Duodenal lesions were scraped, smeared on glass slides, and 16 representative samples from each diet (4 samples per day) were directly plated (in duplicate) onto blood agar and incubated anaerobically at 37°C for 48 h.

A 5-point scoring system used by Gholamiandehkordi et al. (2007) was planned to score intestinal lesions, but in the present experiment, all observed lesions were focal necroses with a size of 1- to 10-mm diameter and the rest of the intestinal sections had no necrotic lesions. Therefore, only 2 scores were used: 0, no lesions and 1, focal necroses (1- to 10-mm diameter). Focal lesions varied from 1- to 2-mm white foci to around 10-mm-diameter mucosal depressions with yellowish green mucoid material laid in the periphery. Whenever such lesions appeared alongside the hemorrhagic lesions, these intestinal sections were recorded as necrotic lesions-positive regardless of the level of the other lesions.

In addition, all of the intestinal sections were scored for hemorrhagic lesions regardless of the incidence of necrotic lesions. Sections with no hemorrhagic lesions were scored as zero, petechial hemorrhages (pinpoint hemorrhagic foci of 1- to 2-mm diameter) of the intestinal mucosa were given score 1, purpura hemorrhages (hemorrhagic foci of 3-mm diameter) as score 2, ecchymotic hemorrhages (blotchy or irregular hemorrhages up to 1 to 2 cm in size) as score 3, coalesced paint brush hemorrhages as score 4, and extravasation as score 5.

Serum Antibodies for α Toxin

The collected blood was allowed to clot for 8 to 10 h at room temperature. The serum was then separated from the coagulum and stored at –20°C. Eight serum samples from each pen were tested for antibody levels developed against the α toxin of *C. perfringens* using an indirect ELISA (Heier et al., 2001; Lovland et al., 2003). Each well of the Nunc immunoplates (F96 Maxisorp, 735-0083, Thermo Fisher Scientific, Rochester, NY) was filled with 100 μL of known antigen, phospholipase C type XIV (Sigma-Aldrich P 4039, St. Louis, MO), 10 μg/mL in sodium carbonate and bicarbonate coating buffer at pH 9.6.

The plates were incubated at 4°C overnight to coat the wells with the antigen. The plates were then washed 3 times with PBS at pH 7.2, with 0.05% Tween 20 (PBST). A volume of 150 μL of 1% BSA in PBST was added to each well as a blocking buffer and incubated at 37°C for 2 h. The plates were then washed again in PBST and left for 1 h at room temperature to dry. The antigen-coated plates were individually sealed, packed, and stored at 4°C.

Serum samples were diluted at 1:250 in PBST and 50 μL of each sample was added into duplicate wells. The plates were left at 4°C overnight and then washed 3 times with PBST. Rabbit anti-chicken immunoglobulin IgY whole molecule conjugated with alkaline phosphatase (Sigma-Aldrich A9171) was diluted (10^{-4}) in blocking buffer and 50 μL of diluted anti-chicken antibodies was added to each well and then the plates were incubated at 37°C. After 2 h, the plates were washed with PBST.

The plates were incubated with 150 μL of para-nitrophenyl phosphate (Sigma-Aldrich P 7998) for 1 h at 37°C. The reaction was stopped with 50 μL of 2 *M* NaOH. The optical density (OD) was read at

405 nm using a microplate reader (Bench Mark 170-6850, Bio-Rad, Hercules, CA). Resulting OD values were pooled for each pen.

Enumeration of C. Perfringens

Clostridium perfringens that were colonised on the mucosal surface of the proximal jejunum were quantified using the BIO K 086-*C. perfringens* antigen detection kit (Bio-X Diagnostics, Jemelle, Belgium) in a double antibody sandwich ELISA (McCourt et al., 2005). Four birds from each replicate pen were used to quantify *C. perfringens* in the mucosa of the jejunum. Gut samples were thawed at room temperature and washed 3 times in PBS. Then the luminal surface of the proximal jejunum was scraped with a sterile surgical blade and half of the scraping was weighed into a microcentrifuge tube and diluted (2×) with dilution buffer (Bio-X Diagnostics) and mixed vigorously before being allowed to settle for 10 min. (The other half of the scraping was retained for coccidia oocyst counts.) Each sample was filled into 2 adjacent wells (one coated with specific monoclonal antibodies for *C. perfringens* and the other coated with non-specific antibodies) of the ELISA plates. The plates were incubated at 21 ± 5°C for 1 h and then washed 3 times with washing solution (Bio-X Diagnostics).

Subsequently, peroxidase-labelled anti-*C. perfringens*-specific monoclonal antibodies were added to each well and then incubated at 21 ± 5°C for 1 h and washed 3 times. Freshly mixed substrate (hydrogen peroxide) and chromogen (tetramethyl benzidine) was then added to each well. After 10 min, 1 *M* phosphoric acid was added to each well and the resulting OD was read at 450 nm using a microplate reader (Bench Mark 170-6850, Bio-Rad). A broth culture of *C perfringens* was used for serial dilutions for plate counts and at the same time for fixation in the dilution buffer (Bio-X Diagnostics) for ELISA (samples for ELISA were immediately frozen). After the plate counts (each dilution was counted 4 times), the ELISA was performed and a standard curve (r = 0.945) was created (Bench Mark 170-6850). This was used to convert the OD values into bacterial counts and the values were pooled for each pen.

Enumeration of Coccidia Oocysts

Eight birds from each replicate pen were sampled for coccidia oocyst counts in the proximal jejunum. Scrapings as taken for the *C. perfringens* counts were diluted (2×) with saturated NaCl and then placed into both sides of a McMaster counting chamber (Chalex Corporation, Wallowa, OR). The number of coccidia oocysts (all species found in the sample) was counted at 10 × 10 magnification. The counts were pooled for each pen before statistical comparison.

Feed Analysis

Dry matter content of the experimental feeds was determined by drying the samples to constant weight at 105°C in an oven. Ash content was determined by AOAC method 942.05 (AOAC International, 2000). Nitrogen was determined with an automatic analyser (Leco FP-528 nitrogen, Leco Corp., St Joseph, MI) by AOAC 968.06 (Dumas method) using EDTA as the standard and the protein content was calculated as nitrogen × 6.25. Gross energy in the feed was determined with an adiabatic bomb calorimeter (model 1261 isoperibol, Parr Instrument Co., Moline, IL) using analytical grade sucrose as the standard. The fat content was determined with the AOAC 920.39 method using a Soxtec 1043 extraction unit. Soluble and insoluble non-starch polysaccharides (NSP) were determined using AOAC methods 991.43 and 985.29 (for soluble and insoluble polysaccharides, resistant starch, and lignin), AACC methods 32-07 and 32-05 (for total dietary fibre), and AACC method 32-21(for soluble dietary fibre; Megazyme K-TDFR assay kit, Megazyme International Ireland Limited, Wicklow).

Total starch contents of the diets were determined with AOAC method 996.11 for high amylose cereal starch and AACC method 76.13 for gelatinised starch (Megazyme K-TSTA assay kit). Trypsin inhibitor activity was determined by measuring the inhibition of milligrams of bovine trypsin for a gram of sample according to the Kakade method as modified by Smith et al. (1980). Crude protein and gross energy were determined in triplicate samples and all the other analyses were performed in duplicate.

Dietary amino acid samples were oxidised with a hydrogen peroxide-formic acid-phenol mixture. Excess oxidation reagent was decomposed with sodium metabisulphite. The oxidised sample was hydrolysed with 6 *M* hydrochloric acid for 24 h. The hydrolysate was adjusted to pH 2.20, centrifuged, and filtered. The amino acids were separated by ion exchange chromatography (Biochrom 20 analyser, Amersham Pharmacia Biotech, Pittsburgh, PA) and determined by reaction with ninhydrin using photometric detection at 570 nm (440 nm for proline).

For tryptophan analysis, the samples were hydrolysed with 4.2 *M* sodium hydroxide for 23 h. The hydrolysate was adjusted to pH 2.20, centrifuged, and filtered. The tryptophan contained in the hydrolysate was then detected by ion exchange chromatography (Biochrom 20 analyser, Amersham Pharmacia Biotech) and determined by reaction with ninhydrin using photometric detection at 570 nm.

Statistical Analysis

The effects of the protein sources on growth performance, serum antibody levels for α toxin, and the number of coccidia oocysts and *C. perfringens* in the proximal jejunum were compared using a randomised block ANOVA (GenStat Release 10.1, Lawes Agricultural Trust, Rothamsted Experimental Station, Harpenden, UK) using the pen as the experimental unit. Individual treatment differences were compared by a protected least significant difference test using a probability of less than 0.05.

The data obtained for the incidence of intestinal necroses, intestinal hemorrhages, liver lesions, percentage of birds that had mucosal *C. perfringens*, and the pen mortality rates were compared using a non-parametric $\div^2$ test.

Results

The proximate nutrient compositions of the experimental diets were approximately similar. However, the potato diet had higher trypsin inhibitor activity and lower crude fat content than that of fish or soy product-containing diets. Trypsin inhibitor activities of potato-, soy-, and fish-based diets were 3.88, 1.51, and 0.82 mg/g, respectively. The soy diet had the highest starch content among 3 diets. The fish diet had a lower soluble NSP content compared with the potato or soy diets. The insoluble NSP content was highest in the potato diet. The potato diet also had higher levels of valine, leucine, tyrosine, and phenylalanine, whereas the fish diet had high levels of glycine, lysine, and methionine. The soy diet had high Glu and Asp levels compared with the potato and fish diets.

The weight gains and feed intakes of the birds fed the potato or fish diets were significantly lower than those fed the soy diets. Weight gain:feed intake ratio and mortality rate were not affected ($P > 0.05$) by dietary treatment. The birds fed the potato diet had a higher ($P < 0.01$) incidence of necrotic lesions in the duodenum and proximal jejunum compared with the birds fed the soy diet. The necrotic lesion incidence in the birds fed the fish diet was intermediate. No necrotic lesions were identified in the mid or distal sections of the small intestine, in any of the treatment groups, although pseudo-membrane formation was observed in a large proportion of birds (data not presented). There was a higher ($P < 0.05$) level of serum antibodies for *C. perfringens* α toxin in birds fed the potato diet compared with the other 2 diets.

Double hemolytic colonies on blood agar were confirmed as *C. perfringens* with biochemical tests and all gram-stained smears predominantly had relatively large (4 to 6 μm long) gram-positive rods with blunt ends existing as singles or up to 4 loosely attached chains or clusters. They were morphologically similar to *C. perfringens* isolated on blood agar plates. Similar bacteria were also attached to damaged villi tips in histological examination of jejunal sections. Fifty-six percent of the birds had *C. perfringens* colonised on the mucosa of the proximal jejunum, but there were no ($\div^2$2 df = 2.14, P = 0.343) treatment differences (66.7, 45.8, and 54.2% for the potato, fish, and soy diets, respectively) in the percentage of birds that had *C. perfringens* colonised on the jejunum or in their counts.

The birds fed potato or fish diets had a higher ($P < 0.001$) proportion of moderate to severe (scores of 3 to 5) duodenal hemorrhages (64 and 77%, respectively) than the birds fed the soy diet (53%). There were no treatment differences in hemorrhagic lesions in the proximal jejunum and the mid small intestine even though there was a relatively high overall incidence. The birds fed the potato diet had a higher proportion ($P < 0.001$) of hemorrhagic lesions than the soy- or fish-fed birds in the distal ileum of the small intestine. The birds fed the potato diets also had a higher ($P < 0.001$) incidence of liver lesions (96%) than those fed the fish (54%) or soy (46%) diets. Mortality tended ($P < 0.1$) to be higher in potato-fed birds. Although none of the birds had clinical coccidiosis, the birds fed the soy diet had a higher ($P < 0.01$) number of coccidia oocysts than those fed the fish diet.

Discussion

The growth performance of the birds in the experiment was comparable with commercial levels (Aviagen, 2007b). A proportion of the birds in all 3 dietary treatment groups had some evidence of the subclinical form of NE. This confirms the finding of Lovland et al. (2003) that a spontaneous *C. perfringens* infection, with some predisposing factors, is sufficient to reproduce subclinical NE in a proportion of birds in relatively large floor-reared flocks. At the beginning of the experimental period, changes in the environment (used litter mixed into fresh litter might have added some coccidia oocysts and *C. perfringens*) with simultaneous dietary changes would have predisposed the birds to subclinical NE, as described by Parish (1961). The mortality percentages found in the experiment were within the ranges reported in subclinical disease and were within the range commonly found in commercial broiler flocks at this age.

The birds fed the potato protein diet, in comparison to the soy diet, had a significantly greater incidence of necrotic lesions in the duodenum and the jejunum. The pseudo-membrane formation (mucosa covered with golden brown to greenish yellow loosely adherent material) that was observed in the distal and middle small intestine suggests that these birds might have had necrotic lesions in the distal and middle small intestine before sampling and they were recovering from these lesions. In addition to intestinal lesions, the birds fed the potato protein diet had a significantly greater incidence of liver lesions, higher serum antibody levels for *C. perfringens* α toxin, and a tendency for a relatively small increase in the flock mortality rate. The proportion of birds that had *C. perfringens* colonised on the jejunal mucosa tended to be higher in the potato diet treatment. There was no evidence of an increased incidence of *C. perfringens* organisms colonised on the gut wall, confirming the findings of McReynolds et al. (2004), who identified no relationship between the incidence of necrotic lesions and the total *C. perfringens* population of the jejunum.

Higher coccidia oocyst counts were found in the soy-fed birds compared with those fed the fish diet. This indicates that the subclinical coccidiosis did not directly contribute to the pathological lesions in the intestine. However, coccidiosis is a factor that predisposes birds to NE and has been identified to have a synergistic relationship with *C. perfringens* during the development of experimental NE. Coccidia multiplication stages in the intestinal epithelium initiate gut mucosal damage but may then be excluded from their proliferation sites when *C. perfringens* colonise the damaged intestinal epithelium further destructing the enterocytes (Williams et al., 2003). Also, the diphtheritic membrane formation could impede the intraluminal dissemination of extracellular coccidial stages.

The 3 diets used in the present experiment had a similar proximate nutrient composition, despite several nutritional differences. Amino acid imbalance has been suggested as a risk factor for subclinical NE. Although there is evidence of an effect of dietary glycine on *C. perfringens* counts in the small intestine and cecum when the birds are orally dosed with the bacterium, in this experiment, there were no major differences in the glycine levels in potato (9.4 g/kg) and soy (9.0 g/kg) diets. Bacterial degradation of aromatic amino acids, such as tyrosine, produces phenolic and aromatic compounds that are toxic to the intestinal epithelial cells; therefore, the higher levels of aromatic amino acids in the potato diet than in the other 2 diets could have contributed to the initial damage of the epithelial cells.

The growth of *C. perfringens* under in vitro conditions needs 11 amino acids. Muhammed et al. (1975) suggested that some amino acids, such as methionine, stimulate the growth of the bacterium, but Dahiya et al. (2007b) found a reduction of *C. perfringens* populations in the cecum and the ileum of the birds fed high Met-containing diets. However, in the present experiment, there was no difference in Met levels between the soy and potato diets (4.9 g/kg), although the fish diet had a slightly higher level of Met (6.2 g/kg). The differences in amino acid balance between the diets do not appear to be correlated to the observed differences in subclinical NE.

Amino acid availability was not determined in the present experiment, but may have differed between the 3 treatment groups. The determined trypsin inhibitor activity was higher in the PPC-based diet than in the soy- and fish-based diet. A high trypsin inhibitor activity in PPC has been found in other studies (Lee et al., 1985). Therefore, it is possible that the PPC-based diet may have had a low protein digestibility and increased the protein entering the distal small intestine and ceca of the chickens when compared with the soy and fish diets. Taciak and Pastuszewska (2007) found that a PPC-based diet with low protein digestibility modified the cecal fermentation of rats, giving higher levels of ammonia and butyrate instead of acetate and propionate in comparison to a soybean meal diet. High cecal butyric acid concentrations have also been observed in birds with subclinical NE.

The protease inhibitors in PPC can vary considerably depending on variety and quality of potatoes used (Jadhav and Kadam, 1998) and also with the method of processing. Heat coagulation at low pH is the method mainly used for industrial purposes and this method of processing can denature the proteins considerably, resulting in a very low nitrogen solubility at low pH (Knorr, 1982). However, the protease inhibitors are less heat-labile, more soluble (at the whole pH range in the intestine), and have lower molecular weights (5 to 25 kDa) when compared with the storage proteins of potato. Potato also contains high molecular weight (85 kDa) protease inhibitors, which can be cleaved into several functional, lower molecular weight fragments by intestinal trypsin (Walsh and Strickland, 1993). The high trypsin inhibitor activity in the PPC-based diet in the present experiment indicates that some protease inhibitor activity remained in PPC.

The dietary protease inhibitors could also have directly influenced the incidence of NE by intensifying the activity of *C. perfringens* toxins. Trypsin inhibitors stabilise functional properties of phospholipase c (α toxin) for longer when they are directly bound with the toxin (Sof'ina and Rakhimov, 1998), which could facilitate the destruction of phospholipid cell membranes. In addition, the high trypsin inhibitor content and some other possible protease inhibitors in the PPC-based diet could have inhibited pancreatic trypsin, which is able to cleave α (phospholipase c) and â 2 (NetB) (Fisher, 2006) toxins to inactivate the toxic activities. These are 2 major toxins in *C. perfringens* toxin type A, which has been mostly implicated in the etiology of NE (Engstrom et al., 2003) Therefore, the higher trypsin inhibitor activity could have aggravated toxin-induced intestinal necroses in the birds fed PPC-based diets than in the birds fed the fish and soy diets.

The glycoalkaloid content of industrial PPC has been estimated to be 257 mg/kg and this exceeds the widely accepted safety limit of 200 mg/kg in humans. Glycoalkaloids are not heat-labile and also have the ability to disrupt cell membranes, which could damage the intestinal epithelium. *Clostridium perfringens* usually colonise on damaged intestinal epithelium (Parish, 1961). Therefore, the glycoalkaloids together with subclinical coccidiosis could have predisposed the *C. perfringens* damage, which subsequently exacerbated leading to necroses of the intestinal mucosa. Dietary potato glycoalkaloids are known to damage the cells in contact tissues such as liver and blood (Smith et al., 1996); therefore, when they accumulate in the liver, glycoalkaloids could have contributed to the high incidence of liver lesions observed in potato diet-fed birds when compared with the birds fed the other 2 diets.

The potato protein diet had a higher level of insoluble NSP but a low soluble NSP content compared with the soy diet. The corn content in the potato diet was higher than in the soy diet. Corn has a considerable content of insoluble NSP (70 g/kg) in the total NSP (76.3 g/kg; Meng and Slominski, 2005), which is probably the reason for the higher NSP level in the PPC-based diet. The published literature is contradictory on the effect of NSP.

The inclusion of cereals with high levels of NSP such as wheat (Branton et al., 1987), barley (Kaldhusdal and Hofshagen, 1992), and rye (Craven, 2000) increases the incidence of NE. Carbohydrase enzyme addition mitigates the negative effects of a *C. perfringens* challenge (Jia et al., 2009). Conversely, Riddell and Kong (1992) could not demonstrate these effects and Branton et al. (1997) concluded that added complex carbohydrates and fibre in broiler rations reduced the intestinal lesions of NE. However, it is unlikely that the relatively small differences in NSP contributed to the incidence of subclinical NE in this experiment.

The oil content of PPC is very low (0.52 g/kg) and this resulted in a low oil content in the potato-based diet. The low oil content in the potato diet could have been a contributory factor for the increased incidence of subclinical NE. High lipid intakes have been associated with an increased bile acid synthesis and excretion (Reddy et al., 1977). The antibacterial effect of bile salts has been demonstrated by Inagaki et al. (2006). However, Knarreborg et al. (2002a) identified *C. perfringens* and *Enterococcus faecium* as the 2 bacterial species that had the highest bile acid hydrolase activity in the small intestinal flora of broiler chickens.

Dietary soy oil has been shown to significantly reduce the population of *C. perfringens* in the ileal microflora of broiler chickens compared with dietary animal fat (Knarreborg et al., 2002b). In this experiment, the oil composition of fish meal would have differed from other 2 diets, but the soy and potato diets had similar sources (corn and soy) of dietary oils. Therefore, the oil composition could not have contributed much to the disease incidence.

In conclusion, this experiment successfully reproduced spontaneously occurring subclinical NE in broiler chickens fed all 3 experimental diets. There was a significant increase in the incidence of subclinical NE in the birds fed potato protein within a nutritionally complete diet in comparison to soy- or fish-based diets. The differences in NE incidence were not consistent with the relatively small differences in amino acid balance or NSP content, but the high trypsin inhibitor activity, low oil content, and possible heat-resistant toxic compounds of the potato protein diet could have contributed to the increased incidence of subclinical NE in the PPC-based diet-fed birds.

Campylobacter Infection

Campylobacter spp. are bacteria that commonly infect a broad range of livestock species, pets and wild animals. In poultry they tend to multiply in large numbers in the hindgut, principally in the caecae. Campylobacters are a significant cause of enteritis in man. Infected poultry are a potential reservoir of this zoonosis.

Campylobacter jejuni is the commonest species found in poultry. All campylobacters are delicate organisms that survive for relatively short periods outside the host unless protected by organic material, biofilm or engulfed by protozoa.

Campylobacter jejuni infection is not currently considered to be pathogenic in poultry though a *Campylobacter*-like organism is considered to be the cause of 'Vibrionic Hepatitis'.

There are indications that plantar pododermatitis, carcase quality and litter quality are better on farms which tend to have *Campylobacter*-negative stock. The reason for this is unclear. It may be that management that favours dry litter reduces the risk of infection and/

or transmission within the flock. There is an annual cycle with increased risk of infection in the summer months in some countries.

Signs:

- None.

Post-mortem lesions:

- None.

Diagnosis

- Isolation of the organism from caecal contents, cloacal swabs or composite faeces. The organism is sensitive to air so swabs should be collected into transport medium and other samples placed in airtight containers with minimal airspace. Samples should be tested as quickly as possible after collection.

Treatment

- Not required on clinical grounds.

Prevention

- In principle, housed poultry can be maintained free of *Campylobacter* infection by consistent application of excellent biosecurity. Key aspects of this include effective sanitation of drinking water, sourcing of water from high quality supplies, avoidance of contact with pets and other farmed species, good hand hygiene by stockmen, and changing of overalls and boots on entering bird areas.

In practice the success of this will also depend upon the degree of environmental contamination by the organism. For this reason it may be difficult to stop the spread of infection between houses once it becomes established in one house.

Many infections are introduced during thinning or other forms of partial depopulation. Insects and rodents may act as a means of transfer of the infection from the general environment into the poultry buildings.

Research is ongoing on the development of vaccines, phage treatments and competitive exclusion approaches, as well as processing plant technologies to reduce carcase contamination.

Candidiasis, Moniliasis, Thrush

A disease of the alimentary tract of chickens, turkeys, and sometimes other birds and mammals, characterised by thickening and white plaques on the mucosa, especially in the crop but sometimes in the proventriculus, intestine and cloaca, and associated with gizzard erosion. The cause is a fungal yeast, *Candida albicans* and the condition is seen worldwide. Morbidity and mortality are usually low.

The route of infection is normally oral and the organism is often present in healthy animals with disease occurring secondary to stress and poor hygiene.

The fungus is resistant to many disinfectants.

Signs:

- Dejection.
- Poor appetite.
- Slow growth.
- Diarrhoea, possibly confused or masked by signs of the primary disease.

Post-mortem lesions:

- White plaques in mouth, oesophagus, crop, occasionally proventriculus and intestine.
- Raised focal lesions may slough into lumen as caseous material.

Diagnosis

- Lesions, histopathology, microscopic examination of a digested smear (heat in 10% potassium hydroxide) to demonstrate the hyphal forms of the yeast in the tissues. Colonies of this fungus appear as white to ivory colour, smooth and with a yeasty smell.

Treatment

- Nystatin (100 ppm in feed) for 7-10 days, copper sulphate (1 kg/tonne feed) for 5 days, or copper sulphate 1gm/2 litre water for 3 days if approved locally.

Prevention

Avoid excessive use of antibiotics and other stressors. Ensure good hygiene, proprionic acid, sodium or calcium proprionate at 1 kg per tonne continually. A finely divided powder of copper sulphate (where approved) at 200gm/tonne continually or to 14-16 weeks in replacement pullets.

Control of *Candida* through drinking water is sometimes practised with chlorination (e.g. Chlorox, sodium hypochlorite) at 5 ppm. This is economical and effective.

It should be repeated periodically. Take care to provide fresh clean feed and water, uncontaminated by fungi.

Cannibalism, Feather Pecking

A complex multifactorial behavioural problem of poultry and game birds seen worldwide. Morbidity is usually low but mortality is high among affected birds. Predisposing factors include overcrowding, excessive light intensity or variation (e.g. through shafts of light in the house), high temperatures, nutritional deficiencies, feed form (mash takes longer to consume than pellets), tenosynovitis and other diseases affecting mobility, boredom, and strain of bird.

Signs:

- Pecking at feet (especially young chicks) and vents (adult layers and turkey poults 8-12 days old), head, face, wings.
- Feather-pulling.

Post-mortem lesions:

- Skin wounding related to particular signs exhibited.
- Generalised anaemia.

Diagnosis

- Age, distribution of lesions, anaemia. Differentiate from bacterial dermatitis, post-mortem cannibalism.

Treatment

- Correct any husbandry problems. Soluble multivitamins and/ or methionine may be of some benefit in some circumstances. Beak trimming may be necessary. If so it should be carried out carefully by trained operators, complying with local regulations and any relevant codes of practice.

Prevention

- Proper density and temperature, low light level, control ectoparasites. Provision of a diet that closely matches the nutritional requirements of the stock concerned.

Cellulitis

Cellulitis is literally an inflammation of connective tissues. It typically occurs between skin and muscles and between muscles and may be an incidental finding in a range of conditions. However its main importance is as a cause of condemnation in meat poultry, particularly broiler chickens.

In the USA it is called 'Inflammatory Process'. The condition is caused by infection of, often minor, skin wounds by particular strains of *E. coli*, which can replicate in the tissues.

Signs:

- Affected flocks tend to have poorer than average productivity and uniformity, but the affected birds are not readily detectable prior to slaughter.

Post-mortem lesions:

- Typically it presents as exudate ranging from liquid and pale cream pus to yellowish solid plaques of caseous material under the skin of the abdomen and/or in the leg. Many affected birds have no other lesions and are reasonably well grown. Many meat inspectors become skilled at detecting subtle differences in skin colour in the affected birds.

Diagnosis

- Typical lesions.

Treatment

- Treatment would not be possible if the problem is identified at a final depletion. If identified at a thinning there may be time for antibacterial treatment to have some benefit for those birds in the early stages of the problem.

Prevention

- Toe scrapes at 15-25 days of age when feather cover is poor are the most likely predisposing factors. Careful flock management with a view to reducing toe wounds has the greatest impact in controlling cellulitis. Routine monitoring of skin damage at about 25 days of age may be helpful in fostering good practices, though most of the birds showing toe scrapes will not go on to develop cellulitis.

Chicken Anaemia

A viral disease of chickens caused by Chicken Anaemia Virus or CAV. Prior to confirmation that it is in fact a virus it was known as Chicken Anaemia Agent or CAA. Mortality is typically 5-10% but may be up to 60% if there are predisposing factors present such as intercurrent disease (Aspergillosis, Gumboro, Inclusion body heptatitis, etc.) or poor management (e.g. poor litter quality). Transmission is usually vertical during sero-conversion of a flock in lay, lateral transmission may result in poor productivity in broilers.

The virus is resistant to pH 2, ether, chloroform, heat (70°C for 1 hour, 80°C for 5 minutes) and many disinfectants even for 2 hours at 37°C. Hypochlorite appears most effective *in vitro*.

Signs:

- Poor growth.
- Pale birds.
- Sudden rise in mortality (usually at 13-16 days of age).
- No clinical signs or effect on egg production or fertility in parent flock during sero-conversion.

Post-mortem lesions:

- Pale bone marrow.
- PCV of 5-15% (normal 27-36%).
- Atrophy of thymus and bursa.
- Discoloured liver and kidney.
- Gangrenous dermatitis on feet, legs wings or neck.
- Acute mycotic pneumonia.

Diagnosis

- Gross lesions, demonstration of ongoing sero-conversion in parent flock, virus may be isolated in lymphoblastoid cell line (MDCC-MSB1).

Treatment

- Good hygiene and management, and control of other diseases as appropriate, may be beneficial. If gangrenous dermatitis is a problem then periodic medication may be required.

Prevention

Live vaccines are available for parents, their degree of attenuation is variable. They should be used at least 6 weeks prior to collecting eggs for incubation. Their use may be restricted to those flocks that have not sero-converted by, say, 15 weeks.

Immunity: there is a good response to field challenge (in birds over 4 weeks of age) and to attenuated live vaccines.

Serology: antibodies develop 3-6 weeks after infection, and may be detected by SN, Elisa, or IFA.

Chlamydiosis, Psittacosis, Ornithosis

An infection of turkeys, ducks, psittacines, pigeons, man, rarely chickens, caused by *Chlamydia psittaci*, a bacterium of highly variable pathogenicity. It is a 'Scheduled Disease' rarely diagnosed in UK, but occurring probably worldwide. Morbidity is 50-80%, mortality 5-40%.

It is transmitted by contact, faecal dust and wild bird carriers, especially pigeons and robins. Egg transmission does not occur.

Elementary bodies are highly resistant and can survive in dried faeces for many months.

Iodophores and formaldehyde are effective disinfecting agents, phenolics are less so.

Intercurrent salmonellosis and, perhaps, other infections may be predisposing factors.

Signs:

- Respiratory signs.
- Greenish-yellow diarrhoea.
- Depression.
- Weakness.
- Inappetance.
- Weight loss.
- Nasal discharge.
- Conjunctivitis.
- Occasional transient ataxia in pigeons.
- Production drops in naive laying flocks

Post-mortem lesions:

- Vascular congestion.
- Wasting.
- Fibrinous pericarditis.
- Airsacculitis.
- Perihepatitis.
- Spleen enlarged and congested, may rupture in pigeons.
- Necrotic foci in liver.
- Fibrinous pneumonia.
- Congested lungs and air sacs in the turkey.

Diagnosis

- History, signs, lesions. Intracytoplasmic inclusions are helpful but confirmation requires demonstration of causal organisms (Giemsa stain, IFA).

  Serology: complement fixation, Elisa and gel diffusion.

  Differentiate from Duck viral hepatitis, Duck septicaemia.

Treatment

- Tetracycline (200-800 ppm in feed for 3-4 weeks) and/or quinolone medication and supervised slaughter.

Prevention

- Biosecurity, exclusion of wild birds.

  Live and inactivated vaccines are protective although the former result in carriers and the latter require several applications.

Chondrodystrophy, Slipped Tendon or Perosis

Caused by deficiency of manganese, choline, zinc, either singly or in combination (although deficiencies of pyridoxine, biotin, folic acid, niacin may also be involved). This condition is seen in chickens, ducks and turkeys. In turkeys it may be an inherited deficiency of galactosamine.

Signs:

- Short legs.
- Lameness.
- Distortion of hock.
- Slipping of Achilles tendon (or perosis).
- Malposition of leg distal to hock.
- In embryos parrot beak, shortened bones.

Post-mortem lesions:

- Shortening and thickening of long bones.
- Tibia and metatarsus bowed.
- Shallow trochlea.
- Lateral slipping of tendon.

Diagnosis

- Lesions, analysis of feed.

  Differentiate from twisted leg, infectious synovitis, rickets, infectious arthritis, ruptured ligaments.

Treatment

- For flock proceed as for prevention, no value to affected bird.

Prevention

- Addition of manganese, choline, vitamins, correct mineral balance.

Coccidiosis of Turkeys

Infection of turkeys with *Eimeria* spp. This disease is not very common in commercially reared turkeys though most turkey growers receive preventative medication for at least part of their lives.

Five species of *Eimeria* have been identified that cause lesions in turkeys, of which two are associated with significant disease effects. *E. meleagrimitis* affects the upper small intestine, while *E. adenoides* affects the caecae and rectum. *E. gallopavonis* and *E. meleagridis* affect the lower small intestine rectum and caecae, while *E. dispersa* is found in the small intestine.

Signs:

- Huddling.
- Weight loss.
- Depression.
- Watery diarrhoea that may occasionally be blood stained or contain clumps of mucus or shed mucosa.
- Tucked appearance, ruffled feathers.

Post-mortem lesions:

- The affected area of intestine shows thickening of the wall and dilation. The contents may be haemorrhagic or be watery with white material shed from the mucosa.

Diagnosis

- Signs, lesions, microscopic exam of scrapings (oocysts, gamonts). Differentiate from necrotic enteritis.

Treatment

- Toltrazuril, Sulphonamides (e.g. Sulphaquinoxaline), Amprolium.

Prevention

- The ionophore coccidiostats lasalocid and monensin are routinely used in turkey growers, typically to 12 weeks of age. Diclazuril is also used for this purpose.

  Dosage levels of ionophores may be critical to efficacy and safety. Exposure of previously unmedicated birds to these compounds can cause toxicity.

  Salinomycin is toxic for turkeys even at very low doses. Avoid use of tiamulin in ionophore treated birds.

Coccidiosis, Caecal, E tenella

This was at one time the commonest type of coccidiosis and is certainly the most easily diagnosed. It is caused by *Eimeria tenella* and results in lesions in the caecum of chickens worldwide. Morbidity is 10-40% and mortality up to 50%. Transmission as for *E. mitis*.

Signs:

- Depression.
- Ruffled feathers.
- Closed eyes.
- Inappetance.
- Diarrhoea, blood in faeces.
- Production less affected than in some of the other forms of coccidiosis.

Post-mortem lesions:

- Petechiae.
- Thickening, ecchymoses, of caecal mucosa.
- Accumulation of varying quantities of blood and caseous necrotic material in the caecum.

Diagnosis

- Signs, lesions, microscopic examination of scrapings. Differentiate from ulcerative enteritis, histomonosis.

Treatment

- Toltrazuril, Sulphonamides, Amprolium, Vitamins A and K in feed or water.

Prevention

- Coccidiostats in feed, vaccination by controlled exposure, hygiene. *E. tenella* is more common when 'straight' ionophore programmes are used. Shuttle programmes with chemicals in the starter diet usually improve control. In some markets the organic arsenical compound 3-Nitro is used as an aid in the control of caecal coccidiosis. Vaccines are used mainly in breeders but increasingly in broilers. Recovered birds have good immunity to the same parasite.

Coccidiosis, E mitis

This condition of chickens, seen worldwide, is caused by the protozoan parasite *Eimeria mitis*, which colonises the small intestine.

The infective agent is found in litter, faeces and on fomites and birds are infected by the oral route with an incubation period of 2-5 days. The disease occurring is proportional to the amount of infective agent ingested. The parasite is moderately resistant in the environment and highly resistant to conventional disinfectants.

Predisposing factors include exposure to faeces and litter conditions that favour development of the parasite (temperature, humidity).

Signs:

- Reduced feed conversion efficiency and weight gain. May predispose to wet litter, secondary bacterial enteritis.

Post-mortem lesions:

- The lesions are minimal and located in the lower small intestine (ileum) which tends to be pale and flaccid with scattered petechiae.

Diagnosis

- Mild lesions, identification of typical small round oocysts and other stages in fresh scrapings from the small intestine.

Treatment

- Not usually treated but susceptible to the products used for other forms of intestinal coccidiosis.

Prevention

- Normally controlled by anticoccidials in feed. May be included in vaccines.

Coccidiosis, E praecox

Infection of chickens with the protozoan parasite *Eimeria praecox* is spread in exactly the same as that with *E. mitis* but is practically non-pathogenic.

Signs:

- Normally asymptomatic but may cause reduced feed efficiency and reduced weight gain, and predispose to other intestinal conditions.

Post-mortem lesions:

- Minimal, but usually excess liquid and mucus in the duodenal loop.
- Severe infection can cause dehydration through excessive fluid loss.
- The cells of the sides of the villi (not tips) are usually parasitised.

Diagnosis

- Identification of characteristic slightly ovoid oocysts in the duodenum in the absence of *E. acervulina* lesions. It has a very short pre-patent period (c. 80 hours).

Treatment

- Not usually treated but susceptible to the products used for other forms of intestinal coccidiosis.

Prevention

- Normally controlled by anticoccidials in feed. Not usually included in vaccines.

Coccidiosis, Ileorectal, E brunetti

A relatively rare form of coccidiosis affecting chickens worldwide caused by *Eimeria brunetti*. Of moderate to high pathogenicity, it is found in the terminal ileum, caecum and rectum. Morbidity and mortality are variable.

Signs:

- Depression.
- Ruffled feathers.
- Closed eyes.
- Inappetance.
- Poor production.
- Diarrhoea, blood in faeces.

Post-mortem lesions:

- Petechiae and thickening of the distal third or more of intestine, extending into caecal tonsils.
- Severe necrotising enteritis.
- Oocysts in caecum and rectum.

Diagnosis

- Signs, lesions, microscopic examination of scrapings. Differentiate from ulcerative enteritis, caecal coccidiosis.

Treatment

- Toltrazuril, Sulphonamides, Amprolium, Vitamins A and K in feed or water.

Prevention

- Coccidiostats in feed, vaccination by controlled exposure, hygiene. This species is not usually included in vaccines for broilers. There is good immunity to the same parasite in recovered birds.

Coccidiosis, Intestinal, of Ducks and Geese

Young ducks and geese may suffer from *Eimeria* spp infection. In the goose *E. anseris* is the most important, while in ducks *Tyzzeria perniciosa* is most pathogenic. *Tyzerria* has eight sporocysts in each oocyst, compared to four per oocyst for *Eimeria*. Coccidiosis occurs only very rarely in commercially reared ducks in the UK.

Signs:

- Sudden death.
- Depression.
- Blood-stained vent.
- Tucked appearance.

Post-mortem lesions:

- Massive haemorrhage in upper small intestine.

Diagnosis

- Signs, lesions, microscopic examination of scrapings (usually few or no oocysts, large number of merozoites). Differentiate from Duck viral hepatitis, Duck viral enteritis, anatipestifer.

Treatment

- Sulphonamides (e.g. Sulphadimidine 30-600gm/100 birds/day, 3 days on, 2 days off, 3 days on), Amprolium, Vitamins A and K in feed or water.

Prevention

- If required coccidiostats could be used in feed, however this is not routinely practised. Hygiene.

Coccidiosis, Kidney

A disease of geese caused by *Eimeria truncata* that can cause high mortality in geese of 3-12 weeks of age, it can also infect Barbary ducks and swans.

Signs:

- Depression.
- Weakness.
- Diarrhoea - faeces tend to be whitish.
- Reduced feed intake.

Post-mortem lesions:

- Enlarged kidneys.
- Kidneys light grey to greyish pink.
- Tiny white foci and petechiae in the kidneys.

Diagnosis

- Lesions, presence of coccidial stages in fresh scrapings of kidney lesions.

Treatment

- Controlled trials of treatments have not been published.

Prevention

- Good Hygiene.

Coccidiosis, Mid-intestinal, E maxima

One of the more common forms of coccidiosis in commercial broilers. Because of the area of intestine affected it tends to have a significant effect on productivity and susceptibility to necrotic enteritis. Caused by *Eimeria maxima*, of moderate to high pathogenicity it is seen worldwide. Morbidity and mortality are variable.

Signs:

- Depression.
- Ruffled feathers.
- Closed eyes.
- Inappetance.
- Poor production.
- Blood or pigment in the faeces.
- Depigmentation of skin and plasma is especially evident in this form of cocccidiosis and this is commercially important in some markets.

Post-mortem lesions:

- Petechiae and thickening of middle third of intestine.
- Poor absorption of nutrients/pigments.
- Mild to severe enteritis, contents often orange in colour, mucosa tends to be pinker than normal.
- This infection is often associated with *E. acervulina* coccidiosis and there may be large numbers of characteristic oocysts in smears.

Diagnosis

- Signs, lesions, microscopic examination of scrapings. Differentiate from necrotic enteritis, non-specific enteritis.

Treatment

- Sulphonamides, Amprolium, Vitamins A and K in feed or water.

Prevention

Coccidiostats in feed, vaccination, hygiene. This is one of the less immunogenic species, commercial vaccines commonly contain more than one strain of *E. maxima*.

Coccidiosis, Mid-intestinal, E necatrix

A highly pathogenic form of coccidiosis, caused by *Eimeria necatrix*, in which the parasite is present in the small intestine and in the caecum. It occurs in chickens worldwide and has variable morbidity but mortality is high in severely affected birds.

Signs:

- Reduced feed consumption.
- Depression.
- Ruffled feathers.
- Closed eyes.
- Inappetance.
- Poor production.
- Diarrhoea, blood in faeces.

Post-mortem lesions:

- Petechiae and thickening, of middle to posterior third or more of small intestine.
- 'Sausage-like' intestine.
- Severe necrotising enteritis.
- Schizonts seen as white spots through the serosa interspersed with petechiae. Deep scrapings necessary to show large schizonts. Oocyts in caecal scrapings.

Diagnosis

- Signs, lesions, microscopic examination of scrapings Differentiate from necrotic enteritis, other types of coccidiosis.

Treatment

- Toltrazuril, Sulphonamides, Amprolium, Vitamins A and K in feed or water.

Prevention

- Coccidiostats in feed, vaccination, hygiene. This is one of the less immunogenic species, commercial vaccines commonly contain more than one strain of *E. maxima*.

Coccidiosis, Upper Intestinal, E acervulina

This is probably the commonest cause of coccidiosis in chickens and occurs worldwide. It is seen in layers and in broilers, both alone and in association with other species of coccidia and is caused by *Eimeria acervulina*, which is moderately pathogenic. Morbidity is variable and mortality low or absent. *Eimeria mivati* is currently considered not to be a valid species distinct from *E. acervulina.*

Signs:

- Depression.
- Ruffled feathers.
- Closed eyes.
- Inappetance.
- Poor production.
- Diarrhoea.
- Depigmentation.

Post-mortem lesions:

- Thickening, and other lesions, restricted to upper third of small intestine - the duodenum and part of the ileum.
- Petechiae.
- White spots or bands in the mucosa. In severe infections they become confluent and cause sloughing of the mucosa.
- Poor absorption of nutrients/pigments.
- A system of assessing the severity of coccidial challenge by attributing a 'score' is often used. A detailed description is beyond the scope of this book. In general terms a score of 0 indicates no lesions and a score of 4 indicates maximal severity of lesion or death. Various publications provide a photographic key to severity of lesion.

Diagnosis

- Signs, lesions, microscopic exam of scrapings. Differentiate from necrotic and non-specific enteritis.

Treatment

- Toltrazuril, Sulphonamides, Amprolium, in feed or water.

Prevention

- Coccidiostats in feed, vaccination by controlled exposure, hygiene. Immunity is quite short lived (about 30 days) in the absence of continued challenge.

Colibacillosis, Colisepticemia

Coli-septicaemia is the commonest infectious disease of farmed poultry. It is most commonly seen following upper respiratory disease (such as Infectious Bronchitis) or Mycoplasmosis. It is frequently associated with immunosuppressive diseases such as Infectious Bursal Disease Virus (Gumboro Disease) in chickens or Haemorrhagic Enteritis in turkeys, or in young birds that are immunologically immature. It is caused by the bacterium *Escherichia coli* and is seen worldwide in chickens, turkeys, etc.

Morbidity varies, mortality is 5-20%. The infectious agent is moderately resistant in the environment, but is susceptible to disinfectants and to temperatures of 80°C.

Infection is by the oral or inhalation routes, and via shell membranes/ yolk/navel, water, fomites, with an incubation period of 3-5 days. Poor navel healing, mucosal damage due to viral infections and immunosuppression are predisposing factors.

Signs:

- Respiratory signs, coughing, sneesing.
- Snick.
- Dejection.
- Reduced appetite.
- Poor growth.
- Omphalitis.

Post-mortem lesions:

- Airsacculitis.
- Pericarditis.
- Perihepatitis.
- Swollen liver and spleen.
- Peritonitis.
- Salpingitis.
- Omphalitis.
- Synovitis.

- Arthritis.
- Enteritis.
- Granulomata in liver and spleen.
- Cellulitis over the abdomen or in the leg.
- Lesions vary from acute to chronic in the various forms of the disease.

Diagnosis

- Isolation, sero-typing, pathology. Aerobic culture yields colonies of 2-5mm on both blood and McConkey agar after 18 hours - most strains are rapidly lactose-fermenting producing brick-red colonies on McConkey agar.

  Differentiate from acute and chronic infections with *Salmonella* spp, other enterobacteria such as *Proteus*, as well as *Pseudomonas*, *Staphylococcus* spp., etc.

Treatment

- Amoxycillin, tetracyclines, neomycin (intestinal activity only), gentamycin or ceftiofur (where hatchery borne), potentiated sulphonamide, flouroquinolones.

Prevention

- Good hygiene in handling of hatching eggs, hatchery hygiene, good sanitation of house, feed and water. Well-nourished embryo and optimal incubation to maximise day-old viability.

Control of predisposing factors and infections (usually by vaccination). Immunity is not well documented though both autogenous and commercial vaccines have been used.

Contact Dermatitis, Hock Burn, Pododermatitis

Contact dermatitis affects skin surfaces which have prolonged contact with litter, the foot pad, rear surface of the hock and, when severe, the breast area. It is seen in growing broiler chickens and turkeys, and in broiler parents.

Contact dermatitis is commonly monitored at processing as a means of measuring 'welfare status'. Some lesions are superficial, whereas others progress to deep ulcers so the size, stage of resolution and severity of lesion is likely to affect the degree of discomfort or pain suffered.

Pododermatitis is often related to high droppings pH, level of soya bean meal in feed (according to some authors, stickiness of droppings), and, most importantly, litter moisture. It can be reproduced by adding water to the litter.

Signs:

- Blackened skin progressing to erosions and fibrosis on the lower surface of the foot pad, at the back of the hocks, and sometimes in the breast area. If severe the foot and hock lesions may contribute to lameness or serve as a portal of entry for secondary infections.

Post-mortem lesions:

- As described under signs.

Diagnosis

- Signs and lesions.

Treatment

- Not applicable.

Prevention

- This condition is closely linked to intestinal function as this is the prime determinent of faecal consistency and stickiness.

  Choice of drinker type (nipple as opposed to bell), drinker management, proper insulation in cold climates, and adequate ventilation to remove moisture are all important.

Cropworms

The nematode worms *Capillaria* spp and *Gongylonema ingluvicola* infect the mucosa of the crop and oesophagous of poultry and game birds. Some have beetle or earthworms as intermediate hosts.

Signs:

- Anaemia.
- Emaciation.

Post-mortem lesions:

- Inflammation and thickening of mucosa of crop and oesophagus.
- White convoluted tracks in the mucosa.

Diagnosis

- Microscopic examination of mucosal scraping.

Treatment

- Levamisole, Coumaphos.

Prevention

- Effective cleaning of housing. Avoidance of access to intermediate hosts. Routine worming.

Cryptosporidiosis

Cryptosporidia are related to the coccidia, but much smaller (typically oocysts are less than ¼ of the size of an *E. acervulina* oocyst). They replicate in the brush border on the surface of epithelial cells.

They also differ from coccidia in being poorly host specific, although bird strains do not infect mammals very well, and vice versa.

Cryptosporidium baileyi can cause respiratory disease in chickens and turkeys.

The same species causes infections of the hindgut and cloacal bursa in chickens, turkeys, and ducks. *C. meleagridis* also infects both species. A further species causes respiratory disease in quail.

The oocysts are excreted ready sporulated in the faeces and infection occurs by inhalation and ingestion.

Signs:

- Snick.
- Cough.
- Swollen sinuses.
- Low weight gain.
- Diarrhoea.

Post-mortem lesions:

- Sinusitis.
- Airsacculitis.
- Pneumonia.

Diagnosis

- Identification of the parasites attached to the epithelium by microscopic examination (smears histopathology acid-fast staining).

Treatment

- Unfortunately there is currently no known effective treatment in poultry. If other disease processes are complicating the situation (e.g. coli-septicaemia) there may be benefit in medicating for these.

Prevention

- The oocysts of cryptosporidia are extremely resistant to chemical disinfection. There are no effective preventative medicines or feed additives. It is becoming increasingly common for water companies to screen water supplies for cryptosporidia because of the human health implications of mammalian strains. Steam cleaning is effective in reducing infection as oocysts are inactivated above about 65°C.

Dactylariosis

A rare fungal disease of chickens and turkeys caused by *Dactylaria gallopava.*

Signs:

- Incoordination.
- Tremors.
- Torticollis.
- Circling, recumbency.

Post-mortem lesions:

- Necrotic lesions with associated congestion in cerebrum.
- Mycotic lesions in lungs, air sacs, etc.

Diagnosis

- Lesions, isolation of the fungus.

Treatment

- None.

Prevention

- Use fresh dry litter (avoid old sawdust).

Degenerative Joint Disease

A condition of chickens and turkeys seen worldwide. The cause remains to be confirmed, but it may result from physical damage, or developmental defects. Morbidity and mortality are low but affected birds are more likely to be 'picked upon' and may end up suffering damage and needing to be culled.

Rapid growth is a possible predisposing factor.

Signs:

- Lameness.
- Reduced breeding performance.

Post-mortem lesions:

- Damaged epiphyseal articular cartilage, especially of femoral anti-trochanter but also other leg joints, resulting in erosions, and cartilage flaps.
- Microscopically there is necrosis and there may be fissures of articular cartilage and associated osteochondritis.

Diagnosis

- Gross and microscopic lesions.

Treatment

- None available.

  Appropriate management of a segregation pen and early marketing of mildly affected birds may limit losses and improve flock welfare.

Prevention

- Avoidance of physical sources of injury to bones and joints. There may be a place for growth-control programmes, especially during period of rapid growth.

Depluming and Scaly Leg Mites

External parasites of adult chickens, pheasants, pigeons etc, *Knemidocoptes* spp.

Signs:

- Cause irritation and the bird pulls feathers.
- Mange lesions on legs and unfeathered parts.
- Unthriftiness.
- Raised thickened scales.

Post-mortem lesions:

- As described under signs.

Diagnosis

- Signs, microscopic examination for mites in scrapings. The adult females are short legged, round, up to 0.5mm in diameter.

Treatment

- Not usually required in commercial poultry. For small flocks dipping the affected parts in a solution of acaricide may be beneficial. Application of mineral or vegetable oil is also beneficial.

Prevention

- Careful cleaning of buildings during down-time will help reduce the risk of these infections. Exclusion of wild birds from chicken areas is advised as far as possible. It may be best to cull affected birds from small flocks.

Dissecting Aneurysm, Aortic Rupture

A complex, genetic condition of turkeys linked to male sex and high growth rate. It has been suggested that degenerative changes in the wall of the artery and copper deficiency may be factors. A sudden noise or other cause of excitement can lead to an 'outbreak', presumably due to a sudden increase in blood pressure.

Signs:

- Sudden death with no warning signs, birds found on breast or side.
- Skin pale.
- Possibly blood in the mouth.

Post-mortem lesions:

- Carcase anaemic.
- Abdominal cavity full of blood.
- Haemorrhages in lungs, kidneys, leg muscles, pericardial sac.
- Rupture of major blood vessel at base of heart or by the kidneys.
- A longtitudinal split of the abdominal aorta is the most common lesion.

Diagnosis

- History and lesions.

Treatment

- None currently licensed. Reserpine, a tranquiliser, was included in feed at 1 ppm for 3-5 days to reduce blood pressure. Aspirin at 250 ppm in feed or water may be of benefit.

Prevention

- Limit feed in birds of 16+ weeks. Reserpine in diet of birds of 4+ weeks (no longer licensed in the UK).

Duck Viral Hepatitis

A viral disease of ducks occurring worldwide and previously a scheduled disease in UK. Morbidity is around 100% and mortality 0-95%.

The disease is transmitted by infected ducks and other waterfowl and spreads rapidly, recovered birds carrying the virus for 8 weeks. The infective agent, a picornavirus may also survive for ten weeks in brooders and five weeks in faeces. A different picornavirus causes a similar condition in North America.

Signs:

- Sudden death.
- Death in good condition.
- Depression.
- Fall on side, paddling of legs, arching of back, rapid deterioration and death, often in opisthotonus.

Post-mortem lesions:

- Liver swollen.
- Punctate/diffuse haemorrhages.
- Kidneys and spleen swollen.
- Microscopically - focal necrosis, bile duct proliferation and inflammation.

Diagnosis

- History, lesions, SN serology, isolation in CE (causes stunting of 9 day embryo). Differentiate from Duck plague (viral enteritis), Duck septicaemia (anatipestifer), coccidiosis, Newcastle disease, Influenza and a 'Type II Variant' hepatitis caused by Astrovirus.

Treatment

- Antiserum, 0.5 ml serum of recovered birds given intramuscularly.

Prevention

- Vaccination and/or antiserum, breeder vaccination. Live, only slightly attenuated vaccine is applied at day old by foot web stab and may be repeated in breeding birds to provide maternal immunity.

Duck Virus Enteritis, Duck Plague

A herpesvirus infection of ducks and geese diagnosed in the UK in 1972, mostly in ornamental collections, in USA since 1967, also the Netherlands and other countries.

All waterfowl are susceptible and the Barbary duck is more susceptible than the Pekin. The disease follows a very acute course with a morbidity of 5-100% and mortality of 5-100%. Transmission is by infected birds, fomites and arthropods. Recovered birds may carry the virus for a year.

Signs:

- Sudden deaths.
- Rapidly spreading disease.
- Drop egg production.
- Photophobia.
- Ataxia.
- Closed eyes.
- Thirst.
- Severe diarrhoea, sometimes dysentery.
- Dehydration.
- Paresis.
- Tremor.
- Occasionally penile prolapse in the penis in drakes.
- Occasionally cyanosis of the bill in the young.

Post-mortem lesions:

- Severe enteritis.
- Crusty plaques from oesophagus to bursa (covered by yellowish plaques in later stages).
- Haemorrhage in intestine, body cavities, heart, pericardium, liver, spleen.
- Young ducks may show thymic and bursal lesions.

Diagnosis

- Isolation: Duck CAMs 12 day embryos die in 4 days, HA-, intranuclear inclusions Differentiate from Duck hepatitis, oesophagitis (birds on restricted feed), vent gleet, pasteurellosis, coccidiosis.

Treatment

- None, but vaccination in face of outbreak is of value, probably through interference.

Prevention

- Isolation from waterfowl, vaccination if approved by authorities (CE adapted live virus).

Dysbacteriosis, Non-Specific Bacterial Enteritis

Inflammation of the small intestine associated with wet litter, excess caecal volume and fermentation, is common in countries with restrictions on the use of antimicrobial growth promoters and pressure to reduce therapeutic antimicrobial usage.

The condition is seen mainly in rapidly growing broiler chickens with good food intake.

Dietary changes, feed interruptions and subclinical coccidiosis may be contributory factors.

No single bacterium appears to be responsible, rather we are dealing with a disruption in the normal flora of the gut.

Signs:

- Diarrhoea.
- Water intake may be increased or irregular.

Post-mortem lesions:

- Excessive fluid content throughout the small intestine.
- Wet faeces in the rectum.
- Voluminous caecae, often with gas bubbles.

Diagnosis

- Signs, lesions, microscopic examination of scrapings from the wall of the small intestine (to exclude coccidiosis and perhaps to assess the bacterial flora).

Treatment

- Amoxycillin and tylosin treatment appear to be beneficial, especially where treatment is initiated early. Treatment should coincide with good relittering and it is important to provide fresh sanitary drinking water.

Prevention

- Competitive exclusion ('normal adult flora') use in day-old chicks reduces the risk of this condition. Feed acidification may be helpful in some circumstances.

  Careful choice of any feed enzymes and their matching with local raw materials can have an impact on substrates made available to intestinal bacteria.

  Good control of coccidiosis. Prophylactic antimicrobial medication may be necessary in some circumstances.

Egg Drop Syndrome, 1976

Egg drop may be defined as a sudden drop in egg production or a failure to achieve a normal peak in production. In the autumn of 1976 a distinct egg drop syndrome was first identified in Northern Ireland. Apparently a similar disease had been seen over a 4-year period in broiler parents in Holland. The cause has been identified as Adenovirus BC14, 127, first isolated in Northern Ireland in 1976.

It affects chickens and has occurred in Ireland, Holland, France, England, Germany, Spain, Peru, Brazil, Uruguay, Argentina. Mortality is usually negligible. Circumstantial evidence suggests that the main route of transmission is through the eggs (vertical transmission) followed by latent infection during rear with viral excretion starting shortly before sexual maturity.

Lateral transmission from bird to bird is slow and may be prevented or slowed for weeks by netting divisions.

Contamination of egg trays at packing stations may play a part in transmission, as may wildfowl and biting insects. Clinical disease occurs during sexual maturity. Spread from house to house may take 5-10 weeks. Unvaccinated flocks with antibodies before lay do not peak normally. The infection is commonly present in ducks and geese but does not cause disease.

Signs:

- Egg drop at peak or failure to peak. Drops may be of 5 to 50% and last for 3-4 weeks.
- Rough, thin or soft-shelled eggs and shell-less eggs.
- Loss of shell pigment.
- Poor internal quality.
- Lack of signs in the birds themselves.

Post-mortem lesions:

- No specific lesion - only a slight atrophy of ovary and oviduct.
- Histopathology - it may be possible to demonstrate degenerative changes in the epithelial cells of the magnum of the oviduct.

Diagnosis

- History, signs/lesions (mainly lack of). Isolation of haemagglutinatin agent in duck eggs or cell culture, group antigen distinct from classical adenoviruses (white cells, throat swabs, oviduct). Serology: HI, SN, DID, Elisa. It is important to rule out other possible reasons for egg drop, which can be caused by a large number of factors acting individually or in combination. Management problems may be involved: inadequate water supply; extremes of temperature; inadequate lighting programme; sudden changes of feed.

Nutritional deficiency should be considered, specifically vitamins E, B_{12}, and D as well as calcium, phosporus, selenium. Diseases in which egg drop occurs, may be infectious or metabolic. Infectious diseases include Infectious Bronchitis, Infectious Laryngotracheitis, Avian Encephalomyelitis, Newcastle disease, Marek's disease/Leukosis or any infectious disease causing a significant systemic disturbance (CRD, Coryza, Cholera, Parasites, Diphtheritic Fowl Pox). Metabolic diseases include Fatty Liver Syndrome, intoxication by sulphonamides, insecticides or nicarbazin.

Treatment

- None. Soluble multivitamins may be recommended as a non-specific measure.

Prevention

- Vaccination with inactivated vaccine prior to lay.

Endocarditis

A condition of chickens associated with several bacterial infections such as staphylococci, streptococci, *Erysipelothrix*, etc.

Signs:

- Fluid-distended abdomen.
- Peripheral vessels congested.

Post-mortem lesions:

- Right ventricular failure and ascites.
- Vegetative lesions usually on the right atrioventricular valve.

Diagnosis

- Differentiate from broiler ascites syndrome by examination of the interior of the heart. Culture of lesions to confirm the bacterium involved.

Treatment

- Medication is only likely to be of value in reducing deterioration in birds that are starting to develop lesions. The choice should depend on sensitivity testing of an isolate.

Prevention

- Good hygiene at turn-around. If successive flocks are affected on the same site prophylactic medication ahead of the anticipated problem may be of benefit.

Epiphysiolysis

A complex condition of chickens that may be associated with trauma, growth plate disease, rickets, bacterial infection, osteomyelitis, and /or trauma.

Signs:

- Apparent dislocation at extremities of long bones.

Post-mortem lesions:

- Separation of epiphyses at the growth plate, sometimes seen in vivo, often occurs or is identified in the processed carcase.

Diagnosis

- Gross inspection, histology may be helpful in identifying underlying problems such as rickets or 'FHN'.

Treatment

- Not applicable.

Prevention

- Control of predisposing factors.

Equine Encephalitis (EEE, WEE, VEE)

A viral disease of pheasants, partridges, wild birds, chickens, turkeys, ducks and pigeons having a high morbidity and high mortality. It is transmitted between birds by pecking and by mosquitoes. These conditions currently only occur from northern South America to North America.

The natural hosts are wild birds and rodents. Horses are also seriously affected.

Signs:

- Nervous symptoms.
- Ataxia.
- Paresis.
- Paralysis.
- Flaccid neck.
- Circling.

- Tremors.
- May also be asymptomatic.

Post-mortem lesions:

- No gross lesions.
- Microscopic lesions not pathognomonic.

Diagnosis

- Isolation in mice, TC, and CE, ID by VN, COFAL.

Treatment

- None.

Prevention

- Protection from mosquitoes, control cannibalism. Vaccinate at 5-6 week.

Chapter 5

Nutrition for Optimising the Immune Response

With high density confinement rearing of livestock, an additional important role of nutrition is that animals are not only fed for production or reproductive performances but must also be fed to minimise infectious disease and their concomitant stresses. In context of Indian Poultry Industry, problem of immunosuppression has been felt to be prominent due to various factors *viz.*, managemental conditions, nutritional status, intensive production system, high density rearing and infectious diseases. Therefore, it is highly essential to find ways and means for enhancement of immune response by nutritional manipulation. Substantial information is available in literature to indicate that administration of certain vitamins, minerals, amino acids and their different combinations to mammals and chicken in excess of their supposed requirements enhances their disease resistance. This increased resistance has been attributed to significant stimulation of humoral and cellular immunity and phagocytosis. Since, the use of antibiotics has been limited, better use of supplementary immuno-stimulatory nutrients has to be made in poultry feeding. Hence, reports of studies undertaken in this direction have been summarised hereunder.

Vitamin A and Zinc

Low and very high dietary vitamin A decreases body weight gain in broilers. Low dietary vitamin A causes depression in *in vitro* T-Lymphocytes responses and *in vitro* antibody production to defined protein antigens. Excess vitamin A intake also decreases immune responses. Maximum T-cell proliferative responses to antigen have been observed at vitamin A levels considerably above NRC (1984) recommended level. Vitamin A is required for intestinal absorption of zinc (Zn) in poultry while zinc influences vitamin A utilisation by affecting retinol binding protein (RBP) synthesis and release from liver. However, the combined level effect of vitamin A and Zn has not been examined on performance and immune response of broilers. Therefore, a study was conducted to find out the effect of supplementation of vitamin A (0, 4500, 8500 and 12500 IU/kg) and Zn (0, 30 and 60 mg/kg) to a basal diet containing vitamin A, 1500 IU/kg and zinc, 35 mg/kg in a factorial experiment design. Chicks were immunised against New Castle Disease Virus (NDV) F-strain by S/C route at day 21. Blood samples were collected at 0, 10 and 21 days of post-immunisation (PI) and serum antibody titers to NDV were determined. For serum antibody titers HI test and ELISA were employed. Optimum growth, best feed efficiency and maximum cellular response was observed in chicks fed diet supplemented with 12500 IU/kg vitamin A and 60-mg/kg zinc. Both HI and ELISA titres were peaked at 10 d post-immunisation on diet supplemented with vitamin A and Zn levels of 12500 IU/kg and 30 mg/kg, respectively.

Vitamin E and Selenium

Vitamin E enhances specific humoral and cell-mediated immune responses as well as native resistance to disease, particularly phagocytosis. Supplemental levels of vitamin E have an immuno-stimulatory effect, increase delayed hypersensitivity and affect mitogenic responsiveness. Dietary supplementation of Se at levels above those recommended as nutritional requirements (0.1 ppm) enhances the primary immune responses. Vitamin E and Se appear to participate in similar nutritional and bio-chemical relationships. Supplementation of vitamin E in the diet of chicks enhances humoral immunity, which may be due to destruction of peroxides by vitamin E. Vitamin E and Se play a role in protecting against oxidative damage. Free radicals are scavenged by vitamin E as a first line of defence and then glutathione peroxidase of which Se is a part destroys any peroxides formed before they can damage the cell. Nutritional deficiencies of vitamin E or Se or both impair immune function as measured by humoral response to sheep red blood cells in young chicks.

Accordingly studies have been conducted with various levels of Se, vitamin E and their combinations to examine the effect on performance and immune response of broilers. Effect of supplementation of vitamin E, selenium and their combinations suggested that maximum body weight gain and best efficiency of feed utilisation were observed in broilers fed diets containing 0.50 mg/kg

Se and 300 IU/kg vitamin E. Significantly higher antibody titres (HI and ELISA) at 10 d PI were attributed to 0.06 mg/kg and 150 IU/kg Se and vitamin E, respectively. Hence, optimum growth and immune response may be achieved at supplemental level of Se of 0.06 mg/kg and vitamin E at 150 IU/kg. The vitamin E level is higher than that of recommended by NRC (1984, 1994).

Methionine and Choline

Maize-soybean diets are first limiting in the sulphur containing amino acids and thus require methionine supplementation. In addition to other functions, methionine serves as donor of methyl groups and a source of sulphur.

Addition of methionine over and above, the recommended requirement of broiler improves their performance in terms of body weight gain and feed conversion efficiency (FCI). Enhanced antibody titre and increased mitogen stimulation by phytohaemaglutinin-P (PHA-P) in chicks have also been reported due to methionine supplementation of maize-soyabean diet. Choline can spare methionine for growth of broiler chicks.

Choline is capable of increasing the FCI to the same degree as L-methionine. Improvement in body weight gain and FCI has also been observed with choline supplementation of chick diet either alone or in combination with methionine. Keeping the above results in view, the present study was undertaken to find out the effect of methionine and choline, either singly or in combination, on the performance and immune response of broilers.

A study was conducted with broilers to find out the effect of addition of methionine (0.0, 1.5, 3.0 and 4.5 g/kg) and choline (0.0, 1000 and 2000 mg/kg) to a basal diet containing crude protein, 221 g/kg; ME, 12.25 MJ/kg; methionine, 3.6 g/kg and choline, 1300 mg/kg. Supplemental levels of methionine and choline were found to be ineffective in improving the growth, food consumption and food conversion efficiency of broilers.

However, a higher value for leucocytes migration inhibition was observed in chicks fed on the diet containing 6.5-g/kg methionine and 1300 mg/kg choline, indicating significantly improved cellular immune response. HI test and ELISA indicated enhanced antibody titres in chicks receiving 3.0 g/kg methionine and 3,300 mg/kg choline, showing a significantly better humoral immune response. Methionine and choline should be supplemental at levels higher than the recommended level of BIS for better health and production in chicks.

Combination of Some Selected Nutrients

Various studies conducted on the effect of supplementation of different combinations of some selected nutrients immune response suggested better cellular and humoral immune responses in broilers fed diets containing combinations different nutrients at levels. Results suggested that supplementation of combination of many above given nutrient Dietary additives for enhancing nutritional value of feeds

Enzyme Supplementation to Improve the Productive Value of High Fibre Diets

Studies conducted on the effect of *in vitro* digestion on the chemical composition of broiler feed incorporated with enzymes suggest an increase in available carbohydrate (A-CHO) and acid soluble nitrogen fraction (ANSF) and reduction in acid detergent fibre (ADF) and neutral detergent fibre (NDF) of low fibre (LF) and high fibre diets supplemented with 1.0 and 1.5 g/kg enzyme feed supplement (EFS). One gram of EFS contained amylase, 7,500 units; cellulase, 400 units; protease, 200 units and lipase, 300 units. LF and HF diets contained un-autoclaved and autoclaved rice bran (RB), wheat bran (WB) and sunflower cake (SC).

The observations have also been supported by *in vivo* studies. These studies involved two types of diets namely LF (fibre, 3.2%) and HF (fibre, 6.4%) containing 22.18% crude protein; 2900 kcal ME and 22.68% crude protein; 2600 kcal/kg ME, respectively. Third diet (HF_1) had 1.5-g/kg enzyme feed supplement. LF diet contained autoclaved rice bran (RB) as higher fibre feed ingredient whereas HF diet contained autoclaved RB, wheat bran and sunflower cake as high fibre feed ingredients. Body weight gain, feed efficiency and performance index were significantly higher in chicks fed diet with EFs. Cost of production of 1 kg live weight of broiler fed LF diet was Rs.13.68, whereas for producing the same live weight gain of broiler fed HF diet with EFS, the cost involved was Rs.13.56.

Therefore, in the production of 1 kg live weight in broiler fed HF diet with EFs, there was saving of 12 paisa approximately. It was also suggested that for large-scale broiler production, use of autoclaved high fibre ingredients along with EFs was economical. The insignificant effect of enzyme supplementation on performance indicates that the enzymes at the concentration used did not elicit any beneficial response on the utilisation of NSP in broilers. The calculated total NSP content in starter and finisher diets is 9.79 and 9.58%, respectively, out of which the major component is glucose, which might be a monomer of cellulose. The observed significant increase in weight gain in xylanase and pectinase fed birds at 28 d of age and subsequent disappearance of this effect in these experiments indicate that the chicken at the younger age are not able to utilise xylans and pectins and thus required additional supplementation of enzymes to hydrolyse these compounds. This observation also indicates that as the age of the bird advances the anti-nutritive effect of these NSP declines.

The growth depression observed in broilers fed xylanase + pectinase + á-galactosidase might be due to incomplete hydrolysis of the NSP and consequent increased viscosity of intestinal digesta. The á-galactosidase concentration employed was very low. The increased viscosity of intestinal digesta is known to inhibit the digestion and subsequent absorption of nutrients from the feed. The growth depressing effect of above enzymes on broilers was alleviated by the addition of cellulase to the above enzyme complex. Since, cellulose is the major component of the NSP in the experimental diets, cellulase supplementation might have increased the hydrolysis of ß-1-4 linkage of glucose monomers utilisation in broilers fed a combination of xylanase + pectinase + á-galactosidase + cellulase. It is also possible that on breakdown of cellulose, the major component of the cell wall the release of encapsulated intracellular nutrients might have been released

Several workers also did not observe any significant effect on broiler performance due to supplementation to corn soya diet of either individual or multi-enzymes hydrolysing NSPs. Contrary to this, a significant improvement in broiler performance has been reported in broilers when fed either corn soya diet or soyabean meal supplemented with either individual or combination of NSP hydrolysing enzymes. The reported beneficial effects of NSP enzymes on broiler performance may be due to increase in digestibility and retention of nitrogen, increase in dry matter and energy utilisation.

The reported wide variation in the performance of broilers due to NSP enzymes supplementation to corn soya diet might be due to variations in the concentration and composition of NSP in the diet and source and/or activity of the enzymes supplemented. The activity of a specific enzyme produced even from the same microorganism vary significantly. Further, the variation could also be due to the variation in dose and composition of enzyme cocktail. The improvement in broiler performance observed by some workers due to supplementation of enzyme could be due to incorporation of higher dose of cellulase (1000 to 3000 IU/kg diet) compared to the levels used in some of other studies (420 and 560 IU/kg diet) and also be earlier workers (15 to 97 IU/kg diet). The lack of response due to supplementation of only xylanase in corn-soya is expected because, the monomers released from by the action of the enzyme on arabinoxylan i.e. arabinose and xylose are poorly metabolised by poultry and therefore of little value to animal performance.

The addition of a mixture of enzymes considering the composition of NSP in a given diet may yield better response compared to supplementation of individual exogenous enzyme. A series of experiments have also been conducted in CARI Experimental Station employing various maize-soya, pearl millet, sorghum, finger millet, mustard cake and un-decorticated sunflower seed meal based diets in broiler chickens, native chickens and Japanese quails. The commercial preparation was analysed and found to contain sufficient enzyme activities. The enzyme preparation was incorporated in diet at level of 50-g/100 kg.

The activities of various enzymes in a gram of commercial enzyme preparation were: ß glucanase, 35732; ß-D xylanopyrosidase, 98466; xylanase, 2202; FT Pase, 397; amylase, 5773 and CM cellulase, 1906 MIU/kg. There was marginal improvement in growth of broiler chicks in pearl millet based diets during starting phase while significant improvement was observed in finger millet based broiler starting and finishing diets. Enzyme supplemented was also found beneficial to improve utilisation of sunflower seed meal supplemented in maize-finger millet or maize-sorghum based broiler starting and finishing diets. However, on repeated trials on maize-soya-deoiled rice bran diets for three times in broiler chickens, twice in Japanese quails and twice in native chickens, enzyme supplementation did not prove beneficial in terms of growth or feed conversion efficiency.

Some of the commonly used feed ingredients for poultry (corn, sorghum, finger millet, deoiled rice bran, soya bean meal, peanut meal, sunflower meal and rapeseed meal) were screened for pentosans, cellulose, pectin and total non-starch polysaccharides. The ingredient *in vitro* digestibilities by enzymes were evaluated. Cereal samples screened contained mainly pentosans. Pectin content was rich in oilseed meals. Sunflower meal, soybean meal, deoiled rice bran and a broiler starter diet were subjected to a 2 stage in vitro digestion assay with 3 different enzyme mixtures *viz.* Enzyme-I (xylanase + cellulase from Trichoderma viridae), Enzyme-II (xylanase + Cellulase + beta-glucanase from Huminicola insolens) and Enzyme-III (xylanase + cellulase + pectinase + beta-glucanase from *Aspergillus aculeatus*) by incubating 0.1 g of the sample with 3 ml of a pepsin-HCl mixture (2000 U pepsin/ml of 0.1 N HCl) for 45 min to simulate the peptic phase of bird digestion. A panereatin-$NaHCO_3$ mixture (2 mg panereatin/ml of 1 M $NaHCO_3$) was used for 2 h at 40°C to simulate the panereatic phase. Digestibility was assessed by measuring the relative viscosity of the digesta supernatant and the total sugars released. Enzyme-I produced the least relative viscosity and highest total sugars in sunflower meal, deoiled rice bran and broiler starter diet, and whereas, Enzyme-III was very effective in soyabean meal subjected to *in vitro* digestion.

Use of Probiotics for Improving the Performance of Birds

Probiotics have been primarily used to establish normal intestinal flora to prevent or minimise the disturbances caused by enteric pathogens and secondarily to serve the function of antibiotic feed additives in diet of animal. The probiotics currently on the market/under investigations contain *Lactobacilli : (L.lactis, L. bulgaricus, L.bifidus, L.brevis, L.cellobiosus, L.fermentum, L.sporogenes, L.acidophillus, L.plantrum, L.cremoris, L.cellinoides, L.salivarius, L.reuteri); Streptococcus (S.faecium, S.lactis, S.thermophilus);* Pdiococcus *(P.halophilus, P.pentosaccus), Bfidobacterium* spp. and Saccharomyces *(S.cerevisine, S.boulardii.), Bacillus (B.cereus, B.subtilis),* etc. These have been used both alone and in combination. However, they are active against a wider range of condition when multiple - strain preparations are used. Lactic acid bacteria or yeast are able to inhibit the growth of bacteria like *Salmonella, Clostridia* and *E.coli*. Probiotics especially lactobacilli and *Bacillus cereus* are also important in the development of immunocompetence against enteric infections.

However, the probiotics are not substitute of antibiotics in birds with serious infections but are useful in restoring the normal bacterial population that was otherwise altered due to administration of antibiotics. The lactic acid bacteria also consume potentially dangerous waste products and suppress the aflatoxin effect.

The response in relation to performance and microbial balance has not always been consistent. Such variations on the response of birds to probiotics feeding have been attributed to several factors such as microbial strain used, number of viable cells per units of feed, mode of administration, processing method for feed employed, type and nature of diet, age of the birds and the sanitary condition of poultry house.

The improvement in the performance of birds has usually been realised for young chicks and more so under the unhygienic housing conditions. Some of the important observations recorded in studies conducted on the use of probiotics are recorded below:

Early establishment of *L.acidophilus,* mannan oligosaccharide and native gut culture in the GIT helps increase the resistance to *Salmonella* colonisation and may even have a possible role in the competitive exclusion by elimination of *Salmonella* organisms from the gut.

The probiotics appear to colonise the GIT of the chicks of a particular strain and also show proper compatibility only with a combination of suitable microbes.

The feeding of probiotics in UV ray-irradiated mash promotes economically more gains in broiler chickens reared under a floor system of management.

Probiotics given at 75 g/100 kg feeds improves growth, feed efficiency, carcass yield and profitability.

An experiment conducted with broiler chicks to evaluate the influence of dietary supplementation of probiotic on immunocompetence, response to *E.coli*, growth and carcass characteristics. The chicks were placed on one of the four dietary treatments. The dietary treatments were a basal diet (control), and three other diets were the same in composition as that of the basal diet but supplemented with probiotic (Probiolac, 100, 150 or 200 mg/kg diet). Supplementation of probiotic 100 mg significantly improved body weight gain (0-4 weeks) but no difference was observed subsequently. Probiotic supplementation of diet did not influence feed consumption or feed conversion ratio. There was significantly higher antibody production in the 100 mg probiotic supplementation group at 10 days and 5 days post-inoculation in response to SRBC antigen when injected at 14 days and 21 days of age, respectively, compared to control. The birds fed probiotic were less susceptible to *E.coli* challenge than controls, however, no difference was observed in the weight of bursa and spleen due to probiotic supplementation. Probiotic had no influence on dressing percentage or weight of internal organs such as liver, heart and gizzard.

Three strains of *Lactobacillus*, NCDC-3 (*L.delbrueckii subsp.lactis*), NCDC-8 (*L.delbrueckii subsp.bulgaricus*) and NCDC-14 (*L.acidophilus*) were offered to 3 groups of 20 chicks. A fourth group served as the control. No differences in 6-week body weights were observed between treatments. Mortality (0-6 weeks) was highest with NCDC-14 and lowest with NCDC-3. However, 9-week body weights were higher in NCDC-3 and NCDC-8 groups compared with the control; higher N, Ca and P balances were observed in these treatment groups. N balance was 0.93, 3.66, 3.28 and 2.70 g, Ca balance was 0.73, 1.06, 0.98 and 0.72 g, and P balance was 0.46, 0.56, 0.51 and 0.39 g, in the control, NCDC-3, NCDC-8 and NCDC-14 groups, respectively. Oral administration of the Lactobacillus strains, NCDC-3 and NCDC-8 to broilers improved growth and feed conversion efficiency.

The influence of Lactobacillus sporogenes in relation to energy on mortality and economics of broiler chicks in summer was investigated. In commercial broilers up to the age of 8 weeks on 2600, 2900 and 3200 kcal ME/kg diet, each with 0.00, 0.02, 0.03 and 0.04% probiotic (*L.sporogenes*) in a 3×4 factorial experiment of completely randomised design. The addition of probiotic decreased the mortality, whereas, progressive increase of ration energy increased the mortality. The economics of raising broilers was assessed by the cost of per kg live weight production. The least cost of production per kg live weight was observed in diet containing 0.02% probiotic and 2600 kcal ME/kg ration.

At a commercial farm in Hissar (Haryana), during winter (January to February) and summer (June to July), broiler chicks were given commercial feeds supplemented or not supplemented (controls) with probiotics (25 g/quintal Lacto-Sacc) for 6 weeks. During the winter, there was a significant ($P<0.001$) reduction in *Escherichia coli* counts (log CFU/g) of caecal contents in Lacto-Sacc fed group (8.00±0.10) compared with controls (9.29±0.08). However, there was no difference in body weight. During the summer, there was a significant ($P<0.001$) reduction in *E.coli* counts (7.99±0.11) and increase ($P<0.05$) in body weight (1402±23.96 g) of the Lacto-Sacc fed group compared with controls (9.31±0.10 log CFU/g and 1317±23.31 g, respectively).

Performance of broiler chicks under different energy and probiotic levels during summer season was observed on commercial broiler chicks fed on diets containing 2600, 2900 or 3200 kcal ME/kg with 0.02, 0.03 or 0.04% probiotic (*L.actobacillus sporogenes*) in a 3×4 factorial experiment of completely randomised design. Weight gain, feed intake and feed efficiency wee recorded. Weight gain was highest for birds given the 2600 kcal ME/kg diet ($P<0.05$), but due to a high feed intake, feed conversion efficiency was lowest ($P<0.05$).

The effect of the probiotics, G.Probiotic, Bioboost-YC and Biospur at 75, 100 and 150% of the level recommended by the manufacturers on body weight gain, feed conversion ratio (FCR), livability and feed costs was studied in 10 treatments involving 460 commercial broiler chicks. After 8 weeks, mean cumulative body weight gain ranged from 1304.89±34.41 (Bioboost-YC, 75 g/tonne, 75%) to 1629.17±40.97 g (Bioboost-YC, 150 g/tonne, 150%), while FCR ranged from 2.46 (Biospur, 375 g/tone, 75%) to 2.83 (control). Mortality did not show any particular trend and ranged from 4.35 to 6.52%. Feed cost per kg live weight gain at the end of 8 weeks, ranged from Rs.13.78 (Biospur, 375 g/tonne, 75%) to Rs.15.73 (control). It is concluded that the use of probiotics in the broiler ration is cost effective and can increase weight gain.

A study on effect of feeding *Streptococcus* culture on the performance of broilers indicated that *S.lactis* supplements in the diet have no beneficial effect on broiler chicks.

Performance of selected strains of *Lactobacillus bulgaricus, Streptococcus* lactis and *Saccharomyces* cerevisiae alone and in different combinations in broilers was investigated. Dietary supplementation of the selected strains of *Lactobaillus bulgaricus* (L4) *Streptococcus lactis* (S1) and *Saccharomyces cerevisiae* (Y3) alone and in different combinations on broilers revealed significant higher growth rates (+14.05 and 10.15%) and feed efficiency ratio (+7.80 and 2.29%) coupled with zero mortality by the combination of the 3 microbes (*L.bulgaricus* + *S.lactis* + *S.cerevisiae*) and L4 of *L.bulgaricus* as compared to control during the starter phase. Contrary to this, none of the treatments showed significantly better performance than control during the finisher phase.

The overall (1 to 6 weeks) growth performance and feed efficiency ratios were significantly ($P<0.05$) better under the treatments involving the combination of the 3 microbes (+7.22 and +2.55%) as well as *L.bulgaricus* (+6.86 and 2.55%) than that of control, and their performance may be correlated to significantly ($P<0.05$) higher digestibilities of dry matter and crude fibre. The differences for the dressing percentage, gut pH and duodenal thickness under different treatments were statistically non-significant. On the basis of overall picture with respect to growth rate, feed efficiency ratio, mortality percent, the combination of the 3 microbes (*L.bulgaricus* + *S.lactis* + *S.cerevisiae*) and L4 strain of *L.bulgaricus* was the most promising probiotics.

A study on effect of probiotic supplementation on growth, nitrogen utilisation and serum cholesterol in broilers was undertaken. In an experiment, broilers were fed for 8 weeks on diets supplemented with probiotic (containing *Lactobacillus acidophilus, L.casei, Bifidobacterium bifidum, Aspergillus oryzae* and *Torulopsis*) 0, 75, 100, 125 mg/kg. Weight gain was 1204.0, 1272.0, 1268.3 and 1210.5 g, respectively. Chickens given probiotic 75 mg retained more ($P<0.01$) nitrogen than controls. Serum cholesterol concentration was lower in probiotic-supplemented birds (93.3 mg/100 ml) than in controls (132.2 mg/100 ml). In an experiment, 80 broilers were fed for 6 weeks on diets supplemented with probiotic 100 mg/kg without or with antibiotic flavophospholipol 100 mg/kg. Probiotic plus antibiotic-supplemented birds had the maximum weight gain (1148.5 g) followed by antibiotic (1141.3 g), probiotic-supplemented (1128.4 g) and control birds (1045.6

g). N retention was greatest in the antibiotic (48.5%) followed by the probiotic (46.5%) and probiotic plus antibiotic-supplemented groups (46.3%) compared with 40.2% in controls. Apparent ME was greatest in birds given probiotic plus antibiotic (12.37 MJ/kg) followed by antibiotic (12.00 MJ/kg), probiotic (11.92 MJ/kg) and controls (11.62 MJ/kg). Serum cholesterol was lower ($P<0.01$) in probiotic-supplemented birds (86.1 mg/100 ml) than in controls (118.4 mg/100 ml).

Prospects of Microbial Phytase in Phosphorus Utilisation by Poultry

Upto about 85% of the total phosphorus found in feedstuffs of vegetable origin, particularly the cereals, cereal by-products and oil cakes is present in form of phytic acid (inositol hexa-hosphate). Under most dietary conditions the phytate phosphorus is poorly utilised by monogastrics including poultry and, consequently, excreted via the feces. The bird's need for phosphorus, therefore, has to be met through supplementing diet with inorganic phosphorus sources such as defluorinated phosphate or the dicalcium phosphate. The traditional practice of adding inorganic phosphorus in poultry ration adds to the cost of feeding besides constraints in finding the right kind of stuff for regular use.

Phytate Problem

Presently in India, about 12 million tones of compounded poultry feed are being manufactured annually. Considering the minimum level of 1% addition in diet, about 0.12 tones of dicalcium phosphate of worth USD400, 000 is required. Besides being an indigestible constituent the phytate phosphorus tends to act as an anti-nutritional factor in diet in that it possesses strong chelating properties thereby markedly reducing the bioavailability of several multivalent cations mainly the Ca^{2+}, Mn^{2+}, Zn^{2+}, Fe^{2+} and Fe^{3+} by forming insoluble, phytate-metal, complexes. Moreover, the phytate also decreased protein digestibility through formation of indigestible protein-phytate complex and strongly inhibit the activity of amylases. Most feed ingredients of plant origin contain from 55-85% of the total phosphorus in phytate form. The cereal by- products like wheat or rice brans and oilseed meals usually possess higher levels of total phosphorus than that in whole grains and with a larger proportion bound as phytate.

Factors Affecting Phosphorus Bioavailability

The practical poultry rations are largely composed of vegetable oil cakes, meals, grains, and their byproducts. Phosphorus bioavailability from such feedstuffs is only about 30%. The hydrolysis and absorption of phytate phosphorus by monogastric animals are complex processes that are affected by certain factors namely- dietary calcium and inorganic phosphorus (or available phosphorus) levels, vitamin D3, age and type (genotype) of birds, types of dietary ingredients, sources of fibre in diet and feed processing etc. The values reported in the literature on the availability in most of the single and many feed mixtures are variable. In feedstuffs of plant origin, the P occurs in association with phytates and the variation in its availability may party be due to the presence of the endogenous phytase which hydrolyses phytate and affects P bioavailability.

High dietary levels of calcium have been reported to have a slightly negative effect on phytate P utilisation in pullets and laying hens. It has been reported that a chick's ability to retain phytate P increases when the diet contained inadequate levels of non-phytin phosphorus (NPP) compared to normal phosphorus levels. Phytate P is more efficiently retained when chickens are fed suboptimal levels of NPP. It is generally conceded that older birds hydrolyse phytate phosphorus to a greater extent than chicks, the basis of this being that there is more of the phytase activity present in the gastrointestinal tract of older birds.

Limited reports indicate that there may be breed and strain differences in the utilisation of phytate phosphorus. Study on the relative ability of three genotype (White Leghorn, Kadaknath and Aseel) on phytate phosphorus utilisation from diets varying in calcium and phosphorus has indicated that the chicks of different genotypes differed in their response to utilise phytate phosphorus.

In view of such undesirable effects of phytic acid, it is preferred to either remove it altogether or reduce its amount in poultry feed or ingredients. Efforts have been made in different laboratories to either eliminate or reduce phytic acid content in plant feedstuffs through chemical methods, solid state fermentation technology, autolysis or by the use of phytase enzyme in diet. Of these, supplementation of diet with microbial phytase appears to be more promising.

In early seventies, Dr.T.J.Nelson and his group at International Minerals and Chemical Corporation, Libertyville, Illinois were among the first to report that inclusion of fungal phytase from *Aspergillus ficuum* in diet of broilers resulted in a marked improvement in the utilisation of phytate phosphorus and also in bone mineralisation. However, because of the high-anticipated production cost of enzyme and a relatively lower cost of di-calcium phosphate or the de-fluorinated rock phosphate, results of their work could not be commercialised.

Making Enzyme to Work

Considering an increased awareness amongst the masses and the environmentalists against high phosphorus in animal wastes in many of the developed countries and a rapid advancement in fermentation technology, interest in application of phytase enzyme in improving phosphorus utilisation by poultry has further been renewed. In this direction, a great number of phytase, hydrolysing microorganisms have been isolated. The strains of *Aspergillus* capable of producing phytase enzyme can greatly meet expectations of the nutritionists. Moreover, steps have been taken towards economic production of phytase through conveying phytase gene in the original wild *Aspergillus* strain on to the production strain thereby resulting in a considerable increase in production efficiency.

Effect of Microbial Phytase on the Performance of Meat Type Birds

Considerable work has been done in India in recent years on the effectiveness of supplemental microbial phytase in improving the P bioavailability in poultry. Microbial phytase supplementation @ 250-500 units/kg in diet increased body weight gain and feed intake in broiler chicken. However, as a result of simultaneous increase in body weight gain and feed intake, several reports have indicated that microbial supplementation had no significant effect on feed to gain ratio in broiler chicken. Improvement in chick performance, however, has been related to an improved utilisation of dietary phosphorus. The phytase supplementation of diet was found associated with a significant increase in plasma inorganic phosphorus concentrations, tibial ash percentages, as well as improved absorption and reduced excretion of dietary phosphorus. However, contrary to these reports, the addition of dietary phytase to corn-soya diet containing less phosphorus than the NRC (1994) recommendation did not improve either body weight gain or feed intake, but it did increase toe and tibia ash and plasma inorganic phosphorus in broiler chickens, growing quails and guinea fowl. The apparent cause of dissimilar results may partly be attributed to the complex nature of factors influencing the bio-efficacy of phytase and utilisation of phytate phosphorus.

Nonetheless, the phytase supplementation to a standard broiler diet can allow the reduction of the usual addition of inorganic phosphate in the diet and also reduce the amount of P excretion into the environment. Enhanced level of vitamin D_3 can also be used as an alternative to phytase to improve phytate availability in low Ca and NPP diets.

Effect of Microbial Phytase on the Performance of Laying Birds

As layer feed differs from broiler feed in respect of among other things, a very high calcium content, the supplemental phytase may just not be as effective in presence of high dietary calcium (3.5%) as in case of broilers (1.0%). Experiments carried out in this direction indicated that phytase supplemented layer diets @ 250-300 units/kg improved laying performance, tibial bone ash and phosphorus absorption.

The phosphorus excretion also decreased significantly in birds fed the phytase-supplemented diet. Quite contrary to these findings. Some workers did not find any improvement in the production performance of laying birds. However, it has been demonstrated that phytase supplementation @ 250 units/kg in layer eliminated inorganic P supplementation without affecting laying performance.

Effects of Microbial Phytase on Mineral Utilisation

Phytic acid has strong chelating potential and forms a variety of complexes with cations and proteins, rendering, these nutrient biologically unavailable. Theoretically, when phytic acid is hydrolysed by microbial phytase, all minerals bound to it should be released.

Relatively, very little work has been done in India on the effect of phytase supplementation on the availability of minerals. Reports on this aspect suggested that contents of ash and minerals in tibia such as Ca, P, Mg, Zn and Fe were significantly higher in phytase supplemented layer diets.

Effects of Phytase on Nutrients Digestibility

Phytase not only improves mineral and trace element utilisation, but may also enhance the protein digestibility and energy utilisation. Some of the recent reports conducted abroad have clearly shown that the supplemental phytase improved apparent ileal digestibility of nitrogen and amino acids (lysine, methionine, cysteine, threonine) and dietary energy.

At the present time, the benefit from adding microbial phytase to poultry diets on the utilisation of minerals and trace elements is undisputed, because many studies conducted abroad and in India have shown consistent effects with differing diets. Thus, it is widely accepted that microbial phytase is an effective means of improving the economics of broiler/layer production. The use of phytase has become accepted all over the world, especially in countries with intensive animal production on limited acreage, where reduction in phosphorus output from animal manure is of ecological concern. Phytase contributes to alleviating excessive P output.

Dietary Manipulation for Enhancing Nutritional Quality of Eggs

Cholesterol is not a fat. It is a waxy fat - like substance produced by all animals including man. Cholesterol is necessary for many bodily functions and serves to insulate nerve fibres, maintain cell walls, produce vitamin D, various hormones and digestive juices. There is little doubt that elevated blood cholesterol levels increase heart attack risk.

Cholesterol content of chicken egg has recently received far more attention than before due to increase in cardiovascular disease in man mainly arthrosclerosis, hypertension and coronary heart disease. Egg yolk is considered one of the richest sources of cholesterol in human diet Cholesterol content of eggs (about 200-250 mg) and blood (around 150mg%) in chickens has found to vary quite considerably. Various researches are presently going on to lower cholesterol content of chicken egg and meat either through the use of additives, dietary fibre, polyunsaturated fatty acid supplementation, etc.

After realising the cardio protective and other beneficial effects of omega-3 fatty acids found in the eggs, the poultry nutritionists have started research to incorporate more of these fatty acids in the egg and have succeeded in developing such an egg called Omega 3 designer egg. They also call it the 'diet egg' or the 'functional egg'.

Diet eggs have a high percentage of Vitamin E, an antioxidant, which prevents oxidation of cholesterol and therefore its ill effects. Cholesterol by itself is not bad. It is harmful only when it is oxidised. These eggs contain 600 mg of Omega – 3 fatty acids. Omega – 3 fatty acids help to reduce cholesterol triglycerides, clog formation, tumour growth and improved immunity. Vitamin 'E' which is a fat-soluble vitamin and is effective antioxidant is enhanced to 100% in Designer Eggs. These eggs are already available in the markets in many countries, including India with good patronage. In order to improve the quality of these eggs further Vitamin E, selenium, and carotenoid pigments are also incorporated into these eggs.

'Diet eggs' produced by a local company M/s Kool Comestibles in Karnataka at Bangalore are priced at Rs.4 each, and come in packs of 12 and 24. World over, there are eleven licensed holders for the technology, who produce and market the diet eggs in 30 countries.

Smart Eggs, which are low in cholesterol and fat. Produced by the Department of Zoology in Punjab University, The first 'diet eggs' in the region are the outcome of research conducted over a period of four years by Prof. S.K.Taneja, Head of PU's Zoology Department. Scoring over the average variety, these eggs have 30 per cent less cholesterol, 300 per cent more vitamin E and double the amount of 'omega3'. Another plus point is that these eggs do not have that marked "eggy" odor. The eggs, priced at Rs.21 for half-a-dozen would be available in leading stores.

Most dietary cholesterol is associated with saturated fats. Dietary cholesterol (1%) and lard (10%) supplementation in Japanese quail was found to be more harmful with increased total serum cholesterol levels and also associated with increased LDL + VLDL cholesterol fractions. Serum cholesterol levels increased with increasing tallow (2-6%) incorporation.

Supplementation of polyunsaturated fatty acids (fish oil or sunflower seed oil, at 2% in diet of laying hens brought significant reduction in serum (20%) and yolk (9%) cholesterol as well as triglycerides (20%). Red palm oil and red palm oil + popped amaranth feeding resulted in a significant reduction in total lipids and cholesterol contents.

Dietary copper at 100 and 150 ppm in diet decreased both plasma and egg yolk cholesterol in laying hens and quails. The dietary guar gum at 0.5 and 1 per cent levels in diet did not alter cholesterol either plasma or egg yolk cholesterol in laying hens and quails.

Onion and garlic were effective as hypocholesterolaemic agents only to limited extent in Japanese quails. Supplementation of Cu at 150 and 250 ppm reduced the serum total cholesterol by about 14% and meat cholesterol by 20%, whereas, garlic at 1% level reduced serum cholesterol by 15% and meat cholesterol by 12% in broiler chicken. Garlic has been found to contain at least 400 constituent compounds of which more than 30 are known to influence the body function minerals such as selenium and germanium, amino-acids such as glutathione, methionine and cysteine, amino alkyl sulfoxides, allicin disulphide oxides, deoxidised supplied and sulphydryl bearing compounds, s-allyl cysteine and the oxide of trithio-doderiens ajoene. Substances like those have been shown to be effective in a wide variety of therapeutic applications.

Serum cholesterol was lower ($P<0.01$) in probiotic-supplemented birds (86.1 mg/100 ml) than in controls (118.4 mg/100 ml). A significant reduction on the total serum cholesterol concentration was observed, but no effect was found on HDL, VLDL, LDL and triglycerides. The addition of probiotic significantly increased the egg production, shell weight, shell thickness and serum calcium, and reduced the concentrations of cholesterol in the serum and yolk.

Choline chloride supplementation had no significant effect on blood glucose, protein and cholesterol levels.

Use of SBM or vegetable protein mixture (33.33% GNC + 33.33% SBM + 33.33% VPMA (VPMA: 37.5% mustard cake + 37.5% till cake + 12.5% cottonseed cake + 12.5% autoclaved linseed meal)) with added lysine and methionine resulted in significantly lower serum cholesterol but higher serum protein content in broiler chicken. No significant ($P<0.05$) differences in blood cholesterol levels were observed on feeding of soybean meal, rapeseed meal or sunflower seed meal in diet of Japanese quail.

A study on effects of diet on egg and serum lipid profile is given below for ready references. White Leghorn hens of 53-58 weeks of age in cages were fed *ad libitum* with designer egg production diet with or without 2g/kg Spirulina, 2g/kg Thulasi *(ocimum sanctum)* leaves or both; along with a control feed. All the four designer egg production diets had 10% linseed (flax), 10% full fat sardine fish having 22% oil and 200mg vitamin E per kg diet. The eggs were collected from 4th week onwards for estimation of their lipid profile.

Utilisation of BT-Cotton, Parental Non-BT Line or Commercial Cotton Seed Meals in Broiler Diet

An effort was made to assess comparative production performance in broiler chickens fed diets containing solvent extracted cottonseed meal (CSM) processed from BT & Parental Non-BT lines. Processed meal of national check and commercial produce cottonseeds were also used for comparison.

The free gossypol contents were 0.02, 0.02, 0.44 and 0.03% in meals of BT, Parental Non-BT, national check and commercial produce cottonseeds, respectively. Day-old broiler chicks (n=243) were divided to 27 groups of 9 each. Nine dietary treatments (iso-nitrogenous, 23% CP and iso-caloric, 2800 kcal ME/kg) were formulated *viz.*, D1 (control, soybean meal-SBM based), D2 and D3 (BT CSM at 10% of diet with and without additional iron), D4 and D5 (non-BT CSM with and without additional iron), D6 and D7 (national check CSM with or without additional iron), and D8 and D9 (commercial produce CSM with or without additional iron at 2ppm for every 1ppm of free gossypol, respectively). Each dietary treatment was offered to three replicated groups upto 6 weeks of age. At the end of 6 weeks of age, 10 birds were taken out randomly from each treatment and were sacrificed to study carcass traits, organs' yield and histo-pathological changes in vital organs.

The broiler chickens received CSM processed from BT (D2, 1753g & D3, 1638g) & Parental Non-BT (D4, 1653g & D5, 1687g) with or without additional Fe grew at same rate as observed in soybean meal (solvent ext.) based diet (D1, 1676g). The feed intake and feed conversion efficiency (feed: gain) in these dietary treatments (BT, non-BT line based diets) also did not differ significantly ($P>0.05$) from control diet. Similar observation was also observed in dietary treatments (D8 & D9) containing solvent extracted cottonseed meal of commercial produce origin. However, a decrease ($P<0.05$) in body weight gain and feed intake was observed in D6 containing national check CSM with high gossypol content. Addition of Fe in the diet (D7) improved ($P<0.05$) feed intake and weight gain but not to the extent as observed in diets containing BT, parental non-BT, and commercial produce CSM or control. However, any type of CSM did not affect feed conversion efficiency when fed with or without additional iron. The carcass characteristics in terms of dressing percentage, liver weight and heart weight was not significantly ($P>0.05$) different between the treatments. The eviscerated yields emanated from diets containing either BT, non-BT or commercial produce were statistically similar to control. However, eviscerated yield of broilers fed national check CSM with or without iron supplementation was lower ($P<0.05$) than BT cotton with Fe supplementation and commercial produce CSM. The study envisaged that BT, parental non-BT and commercial produce solvent extracted cottonseed meal can be included at 10% in soybean meal based broiler diet replacing soybean meal and rice bran without additional iron.

Utilisation of Cotton Seed Meal Prepared from Bollgard-II Cotton with the cry1Ac and cry2Ab2 genes (cry-X gene), Parental Cotton Line or Commercial Cotton in Broiler Diet

A new cotton plant, Bollgard-II cotton, has been produced using biotechnology methods to insert the *cry2Ab2* gene from *Bacillus thuringiensis* var. *kurstaki* into the Bollgard cotton genome. As a result, Bollgard-II cotton produces both the Cry1Ac and the Cry2Ab2 proteins that have insecticidal activity against Lepidopteran insect pests of cotton. The combination of both proteins in one cotton plant provides an additional tool to delay the development of pest resistance to the Cry1Ac protein in Bollgard cotton, because the Cry2Ab2 protein is a different Bt protein class than Cry1Ac. The Bollgard-II traits have been introduced into Indian cotton varieties using traditional plant breeding methods to provide Indian cotton growers an additional tool to protect cotton from lepidopteran insect damage. As a requirement for commercial development of the Bollgard-II product, a study to evaluate the nutritional value of diets containing Bollgard-II cottonseed, parental and commercial lines of cottonseed on chicken performance and carcass yield was conducted in a 42 day feeding study.

The Bollgard-II, parental control and commercial cottonseeds obtained from the Mahyco were first processed with solvent to obtain oil extracted meals. The total and free gossypol (%) contents of cottonseeds were 0.68 and 0.52 in Bollgard-II cottonseeds (MRC 6301 BG II, F_2), 0.64 and 0.47 in parental control cottonseeds (MRC 6301 Non-BG II F_2), 0.63 and 0.49% in the commercial cottonseed variety (NHH-44, F_2).

The total and free gossypol contents were 0.62 and 0.08, 0.54 and 0.05, and 0.61 and 0.07% in meals processed (solvent extraction) from MRC 6301 BG II, MRC 6301 Non-BG II or NHH-44 cottonseeds, respectively. Day-old unsexed broiler chicks (n=280) were divided into 28 groups of 10 each. Seven dietary treatments (iso-nitrogenous, 22% CP and 19.5% CP for 0-3 and 3-6 weeks and iso-caloric, 2900 kcal ME/kg from 0-6 weeks) were formulated *viz.*, D1 (non-cotton control, soybean meal-SBM based), D2 and D3, commercial cotton (NHH-44) CSM at 10% of diet with and without additional iron), D4 and D5 (MRC 6301 BG II cottonseed CSM with and without additional iron), and D6 and D7 (MRC 6301 Non-BG II cottonseed CSM with or without additional iron at 2 ppm for every 1 ppm of free gossypol, respectively). The limiting amino acids lysine, methionine, cystine, threonine and arginine were similar in all the diets in both growth phases (0-3 weeks starting and 3-6 weeks finishing phases).

Each dietary treatment was offered to four replicated groups of birds up to 6 weeks of age. The birds were reared in battery cages with brooding, feeding and watering facility. Other management practices and the vaccination schedule remained similar for all birds. Birds were weighed at weekly intervals with feed intake measured and recorded during the sample interval. A nitrogen retention trial (3 day collection period) was conducted the 6th week. At the end of 6th week, 8 birds per treatment (2 birds/replicate) were sacrificed to study certain blood bio-chemicals, carcass traits, and development of digestive and immune organs.

Body weight gain, feed intake, feed conversion efficiency, and protein and energy utilisation efficiency did not differ statistically ($P<0.05$) either at starting (0-3 wk) or growing (3-6 weeks) and overall (0-6 weeks) growth phase. Dry matter digestibility, nitrogen intake (g/bird/day or per unit energy intake) and nitrogen retention (either expressed as percent of N-intake, g/bird/day or g/unit energy intake) remained statistically similar in all the diets. The different blood biochemical constituents, *viz.* serum protein, albumin and globulin, uric acid, total cholesterol and hemoglobin did not differ ($P<0.05$) across dietary treatments.

The consumption of cottonseed meals did not influence the mortality of birds. Feather loss, blood loss, yield of giblet, gizzard and liver were not different ($P<0.05$) across dietary treatments. However, eviscerated yield in the broilers fed NHH-44 diets was significantly higher ($P<0.01$) than the other treatment groups. All treatment groups had similar eviscerated yield as compared to soybean meal based control except for the MRC 6301 non-BG II + Fe group which was lower and the NHH-44 group which was higher. Also, the yield of heart was statistically higher from the NHH-44 commercial cottonseed meal fed group compared to the soybean meal control, NHH-44 + Fe and MRC 6301 Non-BG II groups. No statistical differences across treatment were observed in the major cut up parts.

Likewise, the development of digestive organs was not affected by dietary treatments except for the weight of the small intestine which was lighter in the birds fed the NHH-44 and NHH-44 + Fe groups as compared to the MRC 6301 BG II, MRC 6301 BG II + Fe and MRC 6301 Non-BGII + Fe groups. The yield of bursa and thymus was significantly higher ($P<0.005$) in MRC 6301 BG II, MRC 6301 BG II + Fe, MRC 6301 Non-BGII, and MRC 6301 Non-BGII + Fe groups as compared to the NHH-44 fed group. Overall, performance, carcass characteristics, nitrogen retention, organ weights and blood biochemical constituents were not different in birds fed cottonseed meal from Bollgard-II cotton compared to the non-transgenic cotton. This indicates that Bollgard-II cotton is as safe and nutritious to chickens as commercial cotton with low free gossypol.

Bio-efficacy of Liquid DL-Methionine Hydroxy Analogue Free Acid (MHA-FA) in Comparison of DL-Methionine in Broiler Chickens

An effort was made to assess the comparative efficacy of liquid DL- methionine hydroxy analogue free acid (MHA-FA) and DL-methionine (MET) in maize-soybean diet on meat production performance of broiler chickens at 6 weeks of age. Three hundred (300) thrifty and bright looking unsexed chicks (commercial strain, Hubbard) were selected, wing banded, weighed and randomly distributed into 30 groups of 10 chicks each.

The experiment was conducted following completely randomised design having 3 dietary treatments with 10 observations (replicates) in each (3 x 10 x 10). Three dietary treatments (iso-nitrogenous, 21% crude protein and iso-caloric, 3150 kcal ME/kg during starting phase i.e. 0-3 weeks; and 20% crude protein with 3200 kcal ME/kg during finishing phase i.e. 3-6 weeks of age) were formulated *viz.*, D_1- Maize-soya based basal diet (Control), D_2- Maize-soya based basal diet+ supplemented with DL- methionine to meet its requirements and D_3- Maize-soya based basal diet supplemented with MHA-FA @ 1.54 times of DL-methionine of treatment 2.

All the birds were housed in battery cages under uniform housing, management and feeding practices except the variations in the dietary treatments as envisaged above. The body weight change of individual birds and feed intake were recorded weekly. Two metabolism trials (total excreta collection method) involving all birds, each of three days duration, were conducted at 3rd and 6th week of age. At the end of 6th week of age, 20 birds were taken out randomly from each treatment and sacrificed to study carcass traits and organs' yield.

The data of 0-3 weeks period revealed that broiler chickens received supplemental methionine (D_2) had significantly higher gain in body weight (472g/bird) which was followed by MHA supplemented group (451 g/b) and un-supplemented control (421 g/b). The live weight gain recorded in MET supplemented group (1095 g/b) during 3-6 weeks of age were similar to that of MHA groups (1092 g/b) but the gains in both the treatments were significantly higher than the un-supplemented control (980 g/b) group. During overall growth phase though highest gain in body weight was recorded in MET group (1570 g/b) but it was statistically similar to MHA group (1543 g/b) and both the treatments resulted in better gain ($P<0.0001$) than un-supplemented control (1400 g/b). The efficacy of methionine hydroxy analogue in comparison to methionine for growth was 62.11, 64.82 and 63.88% at 0-3, 3-6 and 0-6 weeks of age, respectively.

The feed intake during 0-3 weeks of age in the dietary treatment D_2 differed ($P>0.076$) from control diet, but remained similar to that recorded in MHA group. During 3-6 weeks and overall period (0-6wk) the feed intake differed significantly due to treatments being highest in D_2, followed by in D_3 and D_1. Feed conversion efficiency improved significantly on supplementation of either methionine (1.55) or MHA (1.60) in the basal diet containing maize and soybean meal (1.69) during 0-3 weeks of age. During this period the improvement was significantly higher in methionine-supplemented group than MHA group. In contrary, the broilers kept on finisher diet supplemented with MHA (1.92) had better FCR than methionine supplemented (2.00) or un-supplemented group (2.01).

The feed conversion efficiency calculated for overall period however did not differ significantly and the values were 1.98, 1.91 and 1.94, respectively. The efficacy of methionine hydroxy analogue in comparison to methionine for feed efficiency was 62.98, 67.73 and 64.01% at 0-3, 3-6 and 0-6 weeks of age, respectively. The energy and protein efficiency differed during 0-3 weeks of age, while thereafter or during overall growth phase efficiencies remained similar. The mortality (%) recorded were 3, 3 and 4 in D_1, D_2 & D_3 and not attributable to dietary treatments. Dry matter metabolizability did not differ due to dietary treatments. Nitrogen retention increased significantly in diets supplemented with methionine or MHA both at 3rd and 6th weeks of age. Nitrogen retention however, did not differ due to methionine or MHA-FA supplementation.

The carcass characteristics in terms of feather loss ($P<0.011$), eviscerated yield ($P<0.004$), abdominal fat pad ($P<0.007$), and breast yield ($P<0.04$) were significantly different among the treatments. The loss of feather, as percent of live weight, was significantly higher in un-supplemented control (6.16%) than MET (5.20%) or MHA (4.75%) supplemented group. Though feather development did not differ statistically, but it was numerically lower in MHA groups than MET group. The eviscerated yields emanated from diets D3 (72.69 %) or D2 (71.75 %) were statistically higher ($P<0.004$) to that of control (70.09 %). Abdominal fat pad was also significantly higher in broilers fed control diet (2.58 %) than MET (2.23 %) or MHA (2.03 %) supplemented group. Breast yield was higher in MHA (17.34 %) fed broilers than those fed methionine supplemented (16.70 %) or un-supplemented diets (16.14 %).

Crude protein content in breast muscle increased significantly with the addition of methionine or MHA in diet. There was apparent reduction in ether extract contents in breast muscle in MET or MHA group. The composition of thigh muscle did not differ due to dietary treatments. The efficacy of methionine hydroxy analogue in comparison to methionine was 65.85%, 71.40% and 67.49% for eviscerated yield, abdominal fat pad reduction and breast yield, respectively. The weight of spleen as per unit body weight was significantly higher in MET (0.17%) or MHA (0.17%) supplemented diets than control diet (0.13%). The yield of other immune organs (thymus and bursa) were, however, statistically similar. The length of small and large intestine decreased significantly ($P < 0.01$) in MHA supplemented group.

It can be concluded that supplementation of methionine or methionine hydroxy analogue improved growth, efficiency of feed utilisation, eviscerated yield and breast yield. The supposition that methionine hydroxy analogue - free acid (MHA-FA) is about 65% efficient to that of DL- methionine for meat production performance of broiler chickens at 6 weeks of age is valid.

Feasibility of in Ovo Amino Acid Injection for Embryonic Growth and Optimising Total and Digestible Amino Acid Requirements for Meat Production and Immunocompetence of Broiler Chickens

An investigation was undertaken to manipulate the embryonic growth and post hatch gain by *in ovo* injection of amino acids (AA) and dietary supplementation of critical essential AA during successive starting (0-3 wk) and finishing (3-6 wk) growth phases. *In ovo* injection of amino acids with either 22 mm needle at broad end or 11 mm needle at narrow end had no adverse effect on hatchability. *In ovo* injection with 0.5% of total egg amino acids on 14th d of incubation had 73 g higher live weight gain at 6th wk of age as compared to un-injected control. *In ovo* injection of specific Aas i.e. combination of Glycine + Proline, Leucine + Isoleucine + Valine and Threonine + Glycine + Serine AA had higher chick weight, body weight and better immune response at 3rd wk of age. *In ovo* injection of amino acid did not affect the growth of immune organs, digestive organs and percent carcass yield.

The diets formulated on digestible AA had higher body weight and improved feed conversion efficiency than total AA basis. There was no difference in the body weight and FCR of the bird fed 100 and 115% AA of NRC (1994) translated to 3000 kcal ME/kg rather than 3200 kcal as specified by NRC (1994).

The performance of the birds was also not affected by feeding diets with or without animal protein, when the balances of amino acids were met by supplementing the synthetic amino acids. The economic return was higher in low levels of dietary amino acids. Amino acid levels did not affect the immune response and weight of immune organ.

Digestive organ weights were higher in all vegetable meal than fishmeal based diets and in low AA and low ME diets. Cell mediated immune response was higher in low level of AA, while humoral immune response was higher in high energy diets. Low AA and high-energy diets accrued more fat deposition, while breast yield was higher in 115% AA. Breast yield was influenced by AA levels of the diet, whereas ME levels had no effect on the cut-up yields. Nitrogen retention was improved on manipulation of limiting amino acids to 115% of requirement, whereas dietary energy levels did not affect N retention and N loss.

The daily requirement of CP, ME, Lys, Met and Thr for maintenance and live weight gain during starting period were 5.56 g, 216 kcal, 216 mg, 88 mg and 199 mg/$W^{0.75}$ kg and 288 mg, 3.23 kcal, 15 mg, 10.5 mg and 10 mg/g gain, respectively. Based on the weight gain during 0-3 wk period, the requirement of Lys, Met and Thr were predicted as 1.17, 0.75 and 0.86% of the diet, respectively and the ideal ratio of those AA were 100:64:72. The daily requirement of CP, ME, Lys, Met and Thr for maintenance and live weight gain during finishing period were 13.02 g, 129.5 kcal, 588 mg, 238 mg and 699 mg/$W^{0.75}$ kg and 182 mg, 3.92 kcal, 8.6 mg, 3.8 mg and 1.8 mg/g gain, respectively. The requirement of Lys, Met and Thr were predicted as 1.00, 0.42 and 0.67% of the diet, respectively based on the weight gain during 0-3 wk period and the ideal ratio of those AA were 100:42:67. Feeding standard has been developed for both starting and finishing period based on the predicted values, which will be suitable for tropical conditions like India. The study envisaged future scope for identification of specific amino acid(s) manipulating growth of embryo and digestive organs and as immunomodulator and production of designer chicks having nutritive and therapeutic value

Digestibility and Availability of Certain Amino Acids in Soybean Meal and Fishmeal using TME Bioassay

Amino acid rather than total protein requirement is the primary consideration of nutritionists in formulating poultry diets. Recently, it has been suggested that if a ration is formulated on the basis of amino acid (AA) bioavailability instead of AA concentration, it becomes more beneficial for productive performance of birds. For this purpose information on bioavailable contents of AAs in feedstuffs are required. However, at present no such data appears to be available on the feed ingredients of Indian origin. It is more important and pertinent to generate data on bioavailabilities of lysine, methionine and tryptophan since these three amino acids are among the essential and limiting ones in poultry diets. Therefore, studies have been undertaken in order to determine the digestibility and bioavailability of these three AAs in the different samples of commonly used two protein rich feedstuffs viz., SBM and FM in poultry diets.

Samples of soybean (SBM) and fish meal (FM) were collected from across the country and analysed for proximate principles. Ten samples from each feed ingredient were selected on the basis of their variability in respect of chemical constituents. These samples were analysed further for the content of lysine, methionine, tryptophan and FDNB-reactive lysine. Bioassays were conducted with intact and caecectomized roosters for the estimation of digestibility and availability of lysine, methionine and tryptophan as well as apparent (AME) and true metabolizable energy (TME) content in either feedstuff.

The chemical compositions including lysine, methionine and tryptophan concentrations in SBM and FM varied among the samples of both feedstuffs but the magnitude of variation was higher in FM. Biological experiments indicated that lysine was equally digestible (82%) in SBM and FM. However, digestibility of methionine (90%) and tryptophan (87%) in SBM was higher than in (83 and 74%). Bioavailability recorded in SBM for lysine, methionine and tryptophan (85, 93 and 90%) was found to be higher than in FM (83, 83 and 76%). Bioavailability variation in either feedstuff followed a similar trend to that of digestibility. Caecectomy had significant effect only on dry matter digestibility of both feed ingredients but did not exert any influence on the digestibility and bioavailability of nitrogen, three amino acids as well as AME and TME.

SBM had higher nitrogen digestibility (77%) and bioavailability (89%) than that of FM (74 and 88%). AME and TME concentrations of SBM (2.50 and 3.20 kcal/g) were also observed to be higher than that of FM (2.20 and 2.70 kcal/g). The energy metabolizability was highly variable in FM (CV = 25%) as compared to SBM (CV = 2%). FDNB-reactive lysine had high correlation with crude protein, total digestible and bioavailable lysine in SBM or FM. This study also resulted in development of promising prediction equations for estimating the three amino acids content and their bio-availabilities in SBM and FM.

Chapter 6

Toxicity of Certain Inorganic Elements

Current information on toxic dietary levels of inorganic elements for poultry is summarise. A similar summary that describes the mineral tolerances of animals has been provided by the National Research Council (1980b). Toxicity, as defined here, is any adverse effect on performance. Reduced growth rate is the most common criterion used to indicate the specific level at which a particular mineral is toxic. Although most of the information in the table was obtained from experiments in which the mineral was added in the form of an inorganic compound, organic compounds served as the source of minerals in some reports. For instance, some of the information on the toxicity of selenium was obtained by feeding seleniferous wheat.

The toxicity of a mineral is influenced by the nature of the compound in which it is present (for example, methyl mercury is much more toxic than mercuric chloride). Toxicity may also be influenced markedly by the composition of the diet, particularly with respect to other minerals and chelating agents. Selenium included in the diet at 10 ppm reduces the growth rate, but when it is fed in combination with 1,000 ppm of silver, a level as high as 40 ppm does not reduce growth (Jensen, 1975a). Copper at a level of 800 ppm in a practical turkey diet is not toxic, but 50 ppm of copper in a purified diet reduces growth.

The toxicity of copper is modified by the sulfur amino acid content of the diet. Vanadium is much more toxic in a purified diet than in a practical diet, and the toxicity is increased by adding lactose to the practical diet (Hafez and Kratzer, 1976). Conversely, vanadium toxicity is reduced by including cottonseed meal in the diet (Berg, 1965; Berg and Lawrence, 1971; Sell et al., 1986a). In many instances, a high dietary level of one mineral antagonizes another element, resulting in a physiological deficiency of minerals essential for the animal. Because many different factors affect the quantity of a mineral needed to produce toxicity, diverse observations have been reported on the toxic effects of any given mineral.

Composition of Feedstuffs Used in Poultry Diets

Feed formulation involves the judicious use of feed ingredients to supply in adequate amounts and proportions the nutrients required by poultry. Because it is impractical to analyse each batch of feedstuff for its nutrient content, reliance must be placed on feedstuff composition data that have been compiled on the basis of many laboratory analyses. Feedstuffs vary in composition.

The nutrient values given in the following tables are averages reflecting the concentrations of nutrients most likely to be present in the feedstuffs commonly used in poultry feeds.

Feedstuff composition data presented in this edition were obtained from several sources, including the *United States-Canadian Tables of Feed Composition* (National Research Council, 1982), the Association of American Feed Control Officials, commercial firms, and individual scientists. In many instances, the values have been changed to reflect results of analyses of feed ingredients obtained from contemporary crop cultivars and recently employed processing methods.

Additional information provided in the composition tables include nitrogen-corrected true metabolizable energy (TME_n) data for many feed ingredients and information on the true digestibility of amino acids for numerous feedstuffs. Also, equations are provided to estimate the amino acid concentration of certain ingredients on the basis of proximate analysis or on the basis of the protein content of the ingredients.

From a nutritional point of view, there is no "best" diet formula in terms of ingredients that are used. Ingredients should, therefore, be selected on the basis of availability, price, and the quality of the nutrients they contain. Certain ingredients invariably constitute the greatest part of diets, in terms of both amount and cost.

Cereal grains and fats are the primary energy-supplying ingredients, and oilseed meals and animal-protein meals are used commonly as major sources of amino acids. Some important nutritional characteristics of many energy- and protein-supplying ingredients are discussed in this chapter. Sulphur, which are common contaminants in feedstuffs, and their effects are discussed in the final section.

Cereal Grains

Bushel weights (bulk densities) of cereal grains are used in commerce to establish market grades and prices. Bushel weights of grains also have been used as criteria of feeding value, and in some instances this practice seems justified for poultry. For example, at standard moisture levels there is a strong relationship between bushel weight and general feeding value of oats and barley. An increase in bushel weight of these grains is a reflection of an increase in the proportion of the meaty kernel and a decrease in the proportion of fibrous hull. Thus there is a definite increase in the metabolizable energy (*ME*)—and usually protein—content of barley and oats as bushel weight increases. Similarly, there seems to be a direct relationship between the *ME* content of grain sorghum and wheat as bushel weight increases over a wide range.

A relationship between bushel weight and the *ME* content of corn is not so evident. In situations in which corn, sorghum, or wheat fails to achieve maturity because of early frost or early harvest, there usually are decreases in the starchy endosperm portion of the grain and bushel weight and *ME* content are usually low. Regression equations relating the *ME* of corn to various factors such as moisture content at harvest and bushel weight have been reported (Leeson and Summers, 1975, 1976b; Leeson et al., 1977b).

Standard Reference Diets for Chicks

Many laboratories that use Leghorn- or meat-type chicks for studies in animal behaviour, biochemistry, microbiology, nutrition, pathology, physiology, and toxicology need nutritionally complete standard reference diets. The use of standard reference diets that are well defined facilitates more valid comparison of information obtained from experiments conducted within and among laboratories. The diets used successfully in various laboratories and are presented as guides to those requiring such formulations. The isolated soybean protein, casein, and chemically defined diets contain some mineral and vitamin supplements not normally needed in practical diets.

Dextrose (glucose $\cdot H_2O$) rather than starch should be used in diets consisting primarily of purified intact proteins (such as isolated soy protein and casein) to obtain improved performance. Diets containing substantial quantities of dextrose and crystalline amino acids should be stored under refrigeration to minimise Maillard or Browning reactions. These chemically defined diets are intended for short-term use (1 to 3 weeks) and will not support maximum growth over an extended period of time.

Feed Evaluation for Poultry and Rabbits

The intention of this research is to gain a more accurate feed evaluation for poultry and rabbits. This will result in a more efficient nutrient utilisation, a lower environmental pollution and a reduction in the production costs. Feed suppliers as well as feed manufacturers need up-to-date nutritional values for the evaluation of their all-mash animal diets.

The value of a feed concerns different aspects:

- energy and protein values
- amino acid profile
- other nutrients
- macro-minerals
- trace elements
- presence of anti-nutritional factors and
- presence of toxins.

It is important to anticipate the continuously changing quality of feedstuffs. Feed quality is affected by:

- plant selection
- growing
- harvesting techniques
- preservation techniques
- production procedures
- introduction of new or alternative feeds and additives
- new insights in digestion physiology and evaluation techniques.

In Vivo and in Vitro Trials

The nutritional value is determined by means of in vivo digestibility and balance trials, and studies on the caecum-fermentation (mainly in rabbits). The development of in vitro techniques may reduce the use of experimental animals.

The present infrastructure provides 60 balance cages for rabbits and 81 balance cages for poultry (laying hens, broiler chickens and meat turkeys) for the evaluation of feedstuffs, complete diets in order to update our data bank.

Sequential Feeding: a New Perspective in Poultry Feeding

2008 saw the culmination of several years of work on a method of sequential feeding for poultry. This method consists in alternately feeding two diets over 24h to 48h cycles. The sum of the two diets provides the same nutritional balance as that of a single complete diet used in standard feeding.

The interest of the sequential method is that it enables a wider use of raw materials which are difficult to incorporate into traditional feed formulae, and thus it paves the way to possible developments in a context of changing supplies of raw materials.

Use of this method has been shown not to reduce growth or impair zootechnical performance of broilers (Bouvarel et al 2008a) or laying hens (Umar Faruk et al 2008). Trials carried out within the framework of the European Welfare Quality programme have demonstrated that this method can reduce leg problems in fast-growing chickens (Leterrier et al 2008).

Sequential feeding also provides an excellent model for investigating variations in feed intake on a daily basis, and the importance of learning processes and using sensorial cues in feed choice (Bouvarel et al 2008b, 2009).

Finally, this method has enabled the study of short-term regulation of muscle protein metabolism during alternation of diets varying in lysine or in proteins and energy (Tesseraud et al 2009).

Outlook/long-term Impact

Sequential feeding offers a wide range of research possibilities by providing a framework for studying feeding learning paradigms. The rapid changes observed, particularly in the proteolytic pathways, suggest that sequential feeding provides a model that can be easily manipulated for studying chrononutritional phenomena in monogastrics. From an applied perspective, it opens up new feeding methods for poultry by removing the constraints imposed by a single complete diet (use of whole grain cereals, high-protein cereals, etc) and providing a method likely to improve animal welfare. In addition to diversifying rearing conditions, this technique could improve control of growth or enable intramuscular stocks of glycogen to be modulated to improve meat quality.

Partners

These studies have been supported by the European Welfare Quality programme, specific PHASE loans and the 2005-2008 ACTA contract "Alimentation séquentielle et réduction des coûts alimentaires chez le poulet de chair" (Sequential feeding and reduction of broiler feed costs).

The latter involved the Institut Technique de l'AVIculture, poultry feed professionals and researchers at the INRA Centre of Tours.

Forty-Eight-Hour Cycle Sequential Feeding with Diets Varying in Protein and Energy Contents: Adaptation in Broilers at Different Ages

Sequential feeding is a cyclic feeding program with 2 diets for 1 or several days used to induce lower feed costs or to improve welfare quality. The aim of the present study was to investigate the effects of energy [2,800 (E^-) and 3,200 kcal/kg (E+)] and protein [230 (P+) and 150 g/kg (P^-)] content on daily feed intake and growth in 900 male broiler chickens, and to compare these results with standard feeding (CP = 190 g/kg and ME = 3,000 kcal/kg). Sequential feeding was carried out during 48-h cycles in 2 periods (period 1 = 10 to 17 d of age, period 2 = 18 to 29 d of age).

Four treatments were compared during periods 1 and 2: 1) complete diet (C), 2) alternation of diets varying in CP (SP = P+ followed by P^-), 3) in energy (S_E = E^- followed by E+), 4) in protein and energy contents (S_{EPA} = P+E^- followed by P^-E+). A fifth treatment (S_{EPB}) used an alternation in protein and energy contents during period 2 only. All chickens received the same feed during the finishing period (30 to 35 d of age). Feed intake was similar with sequential feeding and complete feed, but in proportion to total feed intake, chickens overconsumed high energy feeds (E+ and E+P^-) during each period, and P^- only for period 2 ($P < 0.01$). During period 2, overconsumption was greater with S_{EPA} than S_{EPB} ($P < 0.01$). Weight gain was similar for all treatments during period 1. At 35 d of age, S_E chickens were heavier than S_{EPA} and S_{EPB} ($P < 0.01$). Feed to gain ratio was similar for all treatments for period 1 and increased for S_P, S_{EPA}, and S_{EPB} compared with C and S_E for period 2 ($P < 0.01$). Walking ability, carcass conformation, breast yield, and abdominal fat did not differ between treatments, but ultimate pH of breast meat was improved with S_P. In conclusion, growth and slaughtering performances similar to standard feeding can be reached with 48-h cycle sequential feeding using diets varying in protein and energy contents.

Sequential feeding is a feeding program, which consists of giving several diets of different nutritional values, for 1- to several-day cycles. It might significantly reduce feed cost and also incorporate higher amounts of cheap raw materials, 1 d out of 2. Several sequential feeding schedules have also been shown to improve welfare by reducing the occurrence of leg abnormalities (Bizeray et al., 2002; Leterrier et al., 2005, 2006).

Sequential feeding with distinct dietary concentrations of energy and amino acids on alternate days has resulted in a similar efficiency compared with a complete feed; this might provide new opportunities to modulate growth (Bouvarel et al., 2004) and to adapt the diet to short-term temperature variations (Lozano et al., 2006).

The first research into sequential feeding dates from the 1970s. Gous and Du Preez (1975) found that when broilers were offered 2 feeds with different amino acid compositions for alternate 6-h periods, the birds were able to adjust their intakes to receive a balanced amino acid intake, but with 12-h alternate feeding there was a deleterious effect on growth rate. Rosebrough et al. (1989) fed broiler chicks high- and low-protein feeds during alternate 24-h periods and observed a decrease in growth.

More recently, Bouvarel et al. (2004) showed in a series of experiments that extreme differences between the composition of the 2 sequential feeds (CP = 300 and 90 g/kg) at 24 h compared with a 48-h cycle of feed distribution had deleterious effects on growth and body composition. In the same studies, experimentations on sequential feeding in 48-h cycles showed that this technique had no negative effect on performance and digestibility of nutrients and very limited effects on muscle and fat deposition in broiler chickens, if the consumption of the sequential feeds provided an energy and amino acid intake similar to the control diet. Lastly, a field trial with 8 flocks of broilers confirmed that feeding high-protein low-energy and low-protein high-energy feeds on alternate days resulted in similar performance to that from feeding a complete feed, despite large day-to-day variations in protein intake. It emerges from these experiments (Bouvarel et al., 2004) that sufficient intake of distinct feeds is essential to reach an overall nutritional balance with sequential feeding.

The present study aimed 1) at evaluating the main effects of energy and protein contents on feed intake during several feeding cycles, 2) to quantify the adaptation of chickens exposed to sequential feeding at different ages, and 3) to quantify time spent in the major behavioural patterns and the occurrence of abnormal gait in the birds. In fact, the occurrence of leg abnormalities may be reduced with alternation of diets varying in energy and protein content (Bouvarel et al., 2004).

Broiler Chickens and Distribution of Feeds

Nine hundred 1-d-old male broiler chicks (Ross PM3) were wing-banded and randomly distributed into 45 pens (1.6 × 1.75 m, 20 chicks per pen, 9 pens per treatment) in an environmentally controlled poultry shed at the Poultry Research Centre in Nouzilly, France. The floor was covered with wood shavings.

Lighting was reduced from daily 24L:0D to 14L:10D after the age of 2 d and was then increased to daily 16L:8D at the age of 7 d. Environmental temperature was progressively reduced from 32 to 23°C, and then maintained at this temperature after the age of 27 d. Feed was provided in linear feeders. All the chickens received the same starter diet (2,900 kcal/kg of ME, 21% CP). From 10 to 29 d of age, chickens were given control or sequential treatments.

Seven diets were used during the sequential feeding period. The control treatment was complete feed (3,000 kcal/kg of ME, 19.5% CP). Two diets were isoenergetic with differences in protein and essential amino acid contents: the high-protein feed (P+ = 23.4% CP) and the low-protein feed (P^- = 15.6% CP). Two diets were isoproteic with differences in ME content: the low-energy feed (E^- = 2,800 kcal/kg of ME) and the high-energy feed (E+ = 3,200 kcal/kg of ME). Two diets had differences in both protein and energy contents: the high-protein and low-energy feed ($P+E^-$) and the low protein and high-energy feed (P^-E+). Diets (i.e., P+ and P^-, E^- and E+, and $P+E^-$ and P^-E+) were formulated to provide on average (50/50) the same nutrient intake as the control feed.

Sequential feeding was carried out during 48-h cycles in 2 periods (period 1: 10 to 17 d of age, period 2: 18 to 29 d of age). Four treatments were compared during periods 1 and 2: 1) complete diet (C), 2) alternation of diets varying in CP (S_P = P+ followed by P^-), 3) in energy (S_E = E^- followed by E+), 4) in protein and energy contents (S_{EPA} = $P+E^-$ followed by P^-E+). A fifth treatment (S_{EPB}) used an alternation in protein and energy contents during period 2 only.

The treatments were arranged in a complete randomised block design with 3 blocks per row and 3 rows. Because Bouvarel et al. (2007) observed a lower feed intake when the cycle started with protein-poor energy-rich diet, protein-rich, energy-poor, or protein-rich energy-poor diets were always given the first day of the cycle to avoid a reduction in the feed intake.

All the chickens received the same feed during the withdrawal period (30 to 35 d of age) (3,050 kcal/kg of ME, 17.5% CP).

Physical Characteristics of Feeds

The diets were manufactured on a semi-industrial scale by the experimental feed mill at the Poultry Research Centre in Nouzilly. They were pelleted using a steam pelleter (Ø = 2.5 mm, L = 47 mm).

One sample of each prepared diet was used for the following measurements. Fine particles were weighed (±0.1 g) after screening 200 g of diet on a 0.6-mm screen. Two replicate measurements were taken for each diet. The percentage of fine particles in the pellets was very low, being 0.6% on average and similar between treatments. Pelleted feed resistance to abrasion was measured using a SABE (SABE Distribution, Chauche, France) durabilimetre with 500-g samples treated for 20 s using a 2-mm screen.

Durability was expressed as the percentage of pelleted particles that resisted abrasion by weighing the fine particles (±0.01 g). Resistance to abrasion for the control feed was 86.0% and ranged from 60.4 to 95.2 for the experimental diets. Hardness was measured in 100 pellets of each diet using an Instron 5543 machine (Instron, Guyancourt Cedex, France).

Hardness was expressed as the maximum load necessary to break the outer surface of the particle (MPa). Hardness of pellets ranged from 0.80 (soft) to 2.13 MPa (hard) for the experimental diets, the control diet was intermediate. Pellet length was measured individually using calipers (200 repetitions/feed).

They varied from 4.1 to 5.2 mm, with the control diet being of an intermediate length (4.5 mm). Colour was measured using a Commission International d'Eclairage L*a*b*c* Hunterlab spectrocolorimetre (Socemi, Metz, France). Mean L* and C* parameters varied from 54.7 to 58.4 and 26.9 to 34.6, respectively. Physical characteristics were dependent on the composition of the feed. A high level of energy obtained by adding oil decreased hardness, resistance to abrasion, length of pellets, and modified colour.

Measurements Performed on Broiler Chicks

Chicken BW was individually measured at 9, 17, 30, and 35 d of age. From 10 to 29 d, feed intake was recorded each day by weighing uneaten feed in each feeder. After exposure to the new feed, and after 2 and 6 h, the percentage of time allocated to 1) eating, 2) standing, 3) pecking, scratching on the floor, or both was measured for the third and seventh cycles. Each pen was observed for 40 s, at 10-min intervals, by 3 different observers.

Each bird was observed in its pen at 34 d of age and assigned into different categories of walking ability by 2 observers, according to a classification adapted from Kestin et al. (1992).

There were 5 categories of gait score, from 0, corresponding to a bird that walked normally with no detectable abnormality, to 4 designating a bird that had a severe gait defect but was still capable of walking with difficulty and only when strongly motivated.

A score of 5 was not included because birds with such a gait score would no longer have been able to walk and would have been culled beforehand. Birds with a gait score of 3 and 4 were considered to have abnormal gait.

At 35 d of age, 4 or 5 chickens per pen whose weights were closest to the pen mean weight were slaughtered (40 broilers per treatment). Breast muscle and abdominal fat weights were recorded.

At 24 h postmortem, the ultimate pH (pHu) of the pectoral muscle was measured with a portable pH metre (model 506, Crison Instruments, Barcelona, Spain) by inserting a glass electrode directly into the thickest part of the pectoralis major muscle.

Statistical Analysis

All data were analysed using STATVIEW program version 5. Tested factors were considered as significantly different if $P < 0.05$. All the data were expressed per pen as an experimental unit.

Nutrient Requirements of Poultry

Poultry diets must be formulated to provide all of the bird's nutrient requirements if optimum growth and production is to be achieved. There are six classes of nutrients:

1. Carbohydrates – the major source of energy for poultry. Most of the carbohydrate in poultry diets is provided by cereal grains.
2. Fats – provide energy and essential fatty acids that are required for some body processes.
3. Proteins – required for the synthesis of body tissue (particularly muscle), physiological molecules (such as enzymes and hormones), feathers and for egg production. Proteins also provide a small amount of energy.
4. Vitamins – organic chemicals (chemicals containing carbon) which help control body processes and are required in small amounts for normal health and growth.

5. Minerals – inorganic chemicals (chemicals not containing carbon) which help control body processes and are required for normal health and growth.
6. Water.

Factors affecting the Nutrient requirements of Poultry

The nutrient requirements of poultry are affected by a large number of factors, including:

- Genetics (the species, breed or strain of bird) – Different species, breeds or strains of bird have different average body sizes, growth rates and production levels and will also absorb and utilise nutrients from feed with different levels of efficiency, leading to different nutrient requirements. As the genetics of commercial poultry is constantly changing, so are their nutrient requirements. Consequently, breeders of commercial poultry provide information on the specific nutrient requirements for the birds they sell.
- Age – Nutrient requirements are related to both body weight and the stage of maturity.
- Sex – Prior to sexual maturity the sexes have only small differences in their nutrient requirements and males and females can usually be fed the same *compromise* diet to achieve acceptable growth rates. Differences in nutrient requirements are larger following the onset of sexual maturity and significantly different diet formulations are then required for each sex.
- Reproductive state – The level of egg production in hens and sexual activity in males will affect nutrient requirements.
- Ambient temperature – Poultry have increased energy requirements to maintain normal body temperature in cold ambient temperatures and the opposite in hot ambient temperatures. The process of digestion of food produces body heat and the amount of heat produced will vary according to the nutrient composition of the diet. This is called the heat increment of the diet. In cold temperatures it may be desirable to formulate a diet with a higher heat increment and the opposite in hot temperatures.
- Housing system – The type of housing system will influence the level of activity of the birds and therefore their energy requirements.

- Health status – Birds experiencing a disease challenge may benefit from an increase in the intake of some nutrients, most commonly vitamins.
- Production aims – Optimal nutrient composition of the diet will vary according to production aims, such as optimising weight gain or carcass composition, egg numbers or egg size. Poultry that are raised for breeding purposes may need to have their energy intake restricted to ensure that they do not become obese.

Nutrient requirements of Meat Chickens (Broilers)

Feed Consumption and Body Weight

There are a number of factors that influence voluntary feed intake. These are discussed in the section on Feed intake. Table below provides data on typical feed consumption and body weight for modern broiler chickens in relation to age and sex.

Table : *Body weight and cumulative feed consumption for male and female broilers (g)*

Age (weeks)	*Male Body weight*	*Male Cumulative Feed Intake*	*Female Body weight*	*Female Cumulative Feed Intake*
0	40	0	40	0
1	170	150	165	145
2	450	480	420	460
3	865	1120	780	1030
4	1410	2020	1250	1825
5	2250	3200	1750	2830
6	2700	4500	2300	4020
7	3350	6000	2800	5400
8	3900	7400	3300	6800
9	4400	8800	3700	8200

Source: Poultry CRC

Nutrient Levels for Broiler Diets

Feeding strategies for broiler chickens will vary depending on the target market for the final product. Strategies for feeding broilers destined for the whole bird market will differ from strategies for broilers destined to be sold as pieces.

Furthermore, the nutrient intake of fast growing broilers must be carefully controlled to prevent metabolic diseases such as ascites and leg weakness. Table below provides data on typical levels of selected nutrients for broiler diets.

Table : *Examples of broiler diets*

Nutrients	*Starter*	*Grower*	*Finisher*
Age fed	0-10	11-24	25-slaughter
Crude protein (%)	22-25	21-23	19-21
ME (MJ/kg)	12.60	13.30	13.50
ME (kcal/kg)	3010	3175	3225
Total Arginine (%)	1.48	131	1.11
Digestible Arginine (%)	1.33	1.18	1.00
Total Lysine (%)	1.44	1.25	1.05
Digestible Lysine (%)	1.27	1.10	0.92
Total Methionine (%)	0.51	0.45	0.39
Digestible Methionine (%)	0.47	0.42	0.36
Total Methionine +Cystine (%)	1.09	0.97	0.83
Digestible Methionine +Cystine (%)	0.94	0.84	0.72
Total Threonine (%)	0.93	0.82	0.71
Digestible Threonine (%)	0.80	0.70	0.61
Total Trypophan (%)	0.25	0.22	0.19
Digestible Tryptophan (%)	0.22	0.19	0.17
Total Valine (%)	1.09	0.96	0.81
Digestible Valine (%)	0.94	0.83	0.70
Calcium (%)	1.0	0.90	0.85
Available phosphorous (%)	0.50	0.45	0.42
Sodium (%)	0.16	0.16	0.16

Source: Ross Broiler Management Manual 2009 (PDF)

Water Consumption Rates for Chickens

Table : *Typical daily water consumption for layers (litres per 1000 birds)*

Production Stage	*Age/Rate of Production*	*21°C*
Layer pullet	4 weeks	100
	12 weeks	160
	18 weeks	200
Laying hens	50% production	220
	90% production	270

Source: Poultry CRC

Table: *Typical daily water consumption for broilers at 20°C (litres per 1000 mixed sex birds)*

Age (weeks)	1	2	3	4	5	6	7	8
Water Intake	65	120	180	245	290	330	355	370

Source: Poultry CRC

Nutrient Requirements of Turkeys

The principles behind formulating turkey diets are similar to those for broiler chickens. Table below provides data on typical nutrient levels for turkey diets.

Table : *Examples of turkey diets*

Nutrients	*Starter1*	*Starter2*	*Grower3*	*Grower4*	*Developer5*	*Finisher6*
Crude protein (%)	28	26	23	21.5	18	16
ME (kcal/kg)	2900	3000	3050	3100	3200	3300
Lysine (%)	1.70	1.60	1.50	1.30	1.15	1.00
Methionine (%)	0.62	0.55	0.50	0.47	0.42	0.34
Calcium (%)	1.4	1.3	1.2	1.3	1.0	0.90
Available phosphorous (%)	0.7	0.6	0.5	0.6	0.5	0.40
Sodium (%)	0.18	0.18	0.17	0.17	0.17	0.17

Source: Commercial Poultry Nutrition, 2nd Ed. S. Leeson and J. D. Summers. Pub University Books, Canada, 1997

Water is an essential nutrient for life. Water consumption can be limited if the water is too hot or is contaminated with excess minerals. Water and food consumption rates are interdependent, so reduced water intake can also lead to reduced food intake. There are other factors that affect water intake, with temperature being the most obvious one. For example, chickens drink between 30-50% more water when the environmental temperature is above 32°C compared with when it is 21°C. Water intake is also affected by the type of drinkers used. The rule of thumb for water intake is that water intake is usually 1.5 to 2 times feed intake. The provide data on typical water consumption levels for layers and broilers, respectively, at 21°C.

Daily Variations in Dietary Lysine Content Alter the Expression of Genes Related to Proteolysis in Chicken Pectoralis major Muscle

Amino acids are major regulators of growth and protein metabolism. For instance, feeding chickens with diets deficient in a specific limiting amino acid (i.e. lysine) decreases growth performance and protein accretion, with a particularly drastic effect observed on breast muscle development (2,3). Lysine supplementation of a lysine-deficient diet otherwise balanced in terms of other amino acids has been shown to greatly modify the amounts of protein synthesized and degraded in chickens (2,4,5).

Similar results have been demonstrated in the growing pig (6) and in humans (7). These findings indicate the importance of optimising animal and human amino acid nutrition, whether quantitative (level of intake) or qualitative (balance between amino acids). The nutritional regulation of protein metabolism also involves another major factor, i.e. the timing of nutrient delivery (1,8). In particular, there is evidence that the timing of protein feeding regulates protein metabolism (9,10). Nevertheless, to our knowledge, little is known regarding the response to temporal or cyclical changes in dietary amino acid composition to date.

Studies performed using lysine-deficient diets have revealed the possibility of a major effect of dietary amino acid levels on protein degradation: the fractional rates of proteolysis (values expressed as %/d) measured in the breast muscle of growing chickens are always higher in lysine-deprived animals (2,4,5), suggesting that the effects of fluctuation in amino acid feeding have to be explored in particular on muscle proteolysis. Protein degradation is a highly controlled, selective, and regulated process.

Three major proteolytic systems are involved in vertebrate muscle proteolysis, i.e. lysosomal, ubiquitin-proteasome–dependent, and Ca^{2+}-dependent systems (11,12). Lysosomal and Ca^{2+}-dependent proteolysis involve cathepsins and the calpain/calpastatin system, respectively. Ubiquitin-proteasome–dependent proteolysis is initiated by the covalent attachment of ubiquitin molecule chains through the successive actions of the ubiquitin-activating enzyme, ubiquitin-conjugating enzymes, and E3 ubiquitin-ligases. Ubiquitinated proteins are then targeted for degradation to the 26S proteasome, which is a large proteolytic complex that hydrolyses protein conjugates.

New evidence has demonstrated that the ubiquitin-proteasome proteolytic pathway is controlled by the expression of 2 important E3 ubiquitin ligases, i.e. Muscle Atrophy F box (MAFbx or atrogin-1)[8] and Muscle Ring Finger-1 (MurF1). These 2 muscle-specific E3 ligases (also called atrogenes) have been shown to be overexpressed in atrophic conditions, including diabetes, cancer, and fasting, and also following disuse (16–21). The aim of this study was to explore the effects of daily variations in dietary lysine supply on lysosomal, Ca^{2+}-dependent, and ubiquitin-proteasome–dependent proteolytic systems in chicken muscle. More precisely, we examined the effect of cyclical changes in dietary lysine in terms of expression of proteolytic genes and change of associated signalling pathways with a focus placed on atrogene regulation.

Animals and Experimental Design

Male chickens (broiler pedigree) were housed in a conventional environmentally controlled poultry shed. The lighting schedule was 24 h of light for the first 3 d of life and a 20-h-light/4-h-dark cycle thereafter (dark period from 0000 to 0400). Environmental temperature was progressively reduced from 32°C to 23°C. Water and food were freely available during the whole experiment period. All the chickens received the same starter feed (metabolizable energy, 12.12 MJ/kg; crude protein, 21%).

From 10 to 24 d of age, chickens were given either control or cyclic feeding programs with alternation of the 2 diets varying in lysine content, each given for 24 h during 48-h cycles as previously described (22). Three experimental diets were used during the sequential feeding period These diets were isoenergetic and isoproteic but contained different lysine contents: control medium-lysine diet (ML) = 100% of the standard lysine level, low-lysine diet (LL) = 70% of the standard lysine level, and high-lysine diet (HL) = 130% of the standard lysine level. To explore the effects of daily variations in dietary lysine supply, sequential feeding programs were carried out by alternating the LL diet with either the ML or HL diet to test possibilities of compensation. In similar conditions, it has been shown that the LL diet was significantly less consumed in the sequence LL/ML but not in the sequence LL/HL (22).

These diets were each given for 24 h and presented in 2 different orders, beginning with either the LL diet during the first period of the 48-h cycle, followed by the HL or ML diet during the 2nd period, or vice versa. The exchange of diets took place at 1400 each day (i.e. the middle of the light period). Body weights were measured throughout the experimental period (on d 10, 17, and 24). At 24 d of age, i.e. after 2 wk of cyclic feeding, chickens were killed at the end of a cycle (between 1400 and 1530) to investigate the impact of sequential feeding on protein metabolism. Five dietary treatments were thus compared: control treatment with the complete standard diet (ML-Ctl); 2 treatments with alternation of LL and HL diets (sequence S1; i.e. HL-S1, chickens receiving the HL diet on d 24, and LL-S1, chickens receiving the LL diet on d 24); and 2 treatments with alternation of LL and ML diets (sequence S2; i.e. ML-S2, chickens receiving the ML diet on d 24, and LL-S2, chickens receiving the LL diet on d 24). In these treatments, the indicated diet corresponded to the last diet in a cycle before tissue sampling. To further support our work conclusively, we also used another model of sequential feeding recently studied by our group as an alternative nutritional strategy for poultry production. Briefly, sequential feeding was conducted during 48-h cycles from 10 d of age with an alternation of diets varying in protein content (low- and high-protein diets) and compared with a control treatment with a complete standard medium-protein diet. For the 2 experiment paradigms, chickens were killed by decapitation after blood sampling. The Pectoralis major muscles were removed and quickly frozen. Frozen tissues were ground in liquid nitrogen and stored at $^{-}80°C$ until analysis. All experiments were carried out with due regard to legislation governing the ethical treatment of animals and investigators were certified by the French government to conduct animal experiments.

Measurement of Muscle Protein, RNA, and Ribosomal Capacity

Frozen powdered muscle samples were homogenized in 2% $HClO_4$ according to the method of Schmidt-Thannhauser as modified by Munro and Fleck (24). Protein content was measured using the bicinchoninic acid kit (Interchim). Total RNA was measured on the basis of the UV absorbance at 260 nm, with a correction for peptide material based on the UV absorbance at 232 nm. Ribosomal capacity (Cs), i.e. the capacity for protein synthesis, was calculated as the ratio of RNA:protein (25).

RNA isolation and RT-PCR

Total RNA was extracted using RNA Now (Biogentec) from 100-mg muscle samples according to the manufacturer's recommendations. After RNase-Free DNase treatment, RNA was reverse transcribed using Super Script II RNase H Reverse Transcriptase (Invitrogen) in the presence of Random Primers (Promega). Quantitative PCR were performed in duplicate using the Abi Prism 7000 apparatus (Applied Biosystems) as previously described (26). The sequences of primers used, specifically designed or reproduced from the literature (27). Gene expression levels were estimated on the basis of PCR efficiency and threshold cycle deviation of an unknown sample compared with a control, as previously described (26). 18S ribosomal RNA was chosen as the reference gene.

Statistical Analysis

Values are expressed as means ± SEM. Statistical analysis was performed using ANOVA (Stateview Software program; SAS Institute) to detect significant intergroup differences. The means were compared by Fisher's least significant difference test in the case of a significant effect and $P < 0.05$ was considered significant. To test the potential relationships between atrogene mRNA levels and protein phosphorylation status, Pearson correlation coefficients were analysed by a Fisher's are to z test.

Growth Performance and Muscle Characteristics

Sequential feeding significantly affected growth performance; within-program body weights from sequential chickens were significantly reduced compared with ML-Ctl chickens ($^{-}$9 and $^{-}$12% for sequences S1 and S2, respectively). Groups, i.e. HL-S1, LL-S1, ML-S2, and LL-S2, did not significantly differ Protein content was significantly lower in the Pectoralis major muscle of chickens fed the LL diet compared with the HL-S1 and ML-Ctl groups, but this result was probably not due to reduced muscle RNA content or Cs (i.e. capacity for muscle protein synthesis). Moreover, feeding an alternation of diets varying in lysine content did not drastically modify the expression of adult fast myosin heavy chain (MyHC), which is the major MyHC in the Pectoralis major muscle of chickens and represents the bulk of muscle proteins.

Muscle Expression of Genes related to Proteolysis

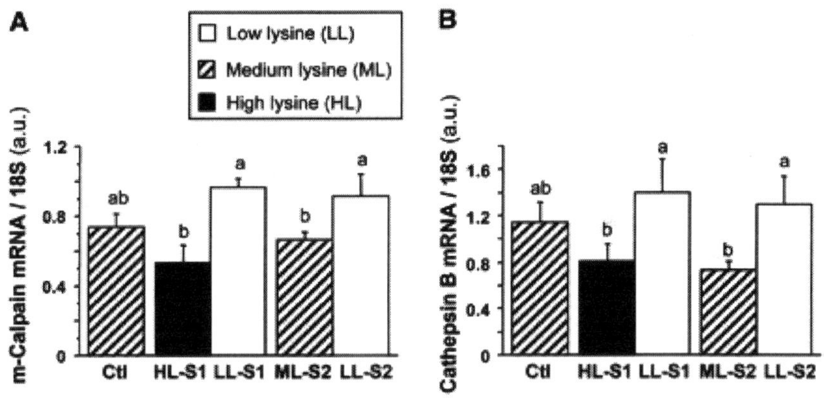

Figure: *Effects of 48-h–cycle sequential feeding with diets varying in lysine content on the expression of genes related to* Ca^{2+}*-dependent* (A) *and lysosomal* (B) *proteolysis in breast muscle of 24-d-old chickens. Analysis was using real-time RT-PCR for m-calpain, cathepsin B, 20S proteasome C2 subunit, and ubiquitin. mRNA levels were normalised using 18S rRNA; 18S was not significantly affected by dietary treatments. Data are expressed as means ± SEM,* n = *6. Means without a common letter differ,* $P < 0.05$*. Whatever the dietary treatment within a sequence, the indicated diet corresponds to the last diet in a cycle before tissue sampling.*

Lysine intake greatly affects muscle protein breakdown, particularly in Pectoralis major muscle (4,5). We therefore investigated muscle expression of genes related to proteolysis. Expression of m-calpain and cathepsin B was significantly affected by dietary treatments, with significantly higher mRNA levels in the muscle of chickens fed the LL diet than in the HL-S1 and ML-S2 groups. The ML-Ctl group exhibited intermediate values, which did not significantly differ from those of the other groups, i.e. HL-S1, LL-S1, ML-S2, and LL-S2.

Conversely, the mRNA levels of the 20S proteasome C2 subunit and ubiquitin were unchanged, irrespective of dietary treatment (data not shown). Because the ubiquitin-proteasome proteolytic pathway is controlled by the expression of E3 ubiquitin ligases, we then examined the potential changes in MurF1 and atrogin-1.

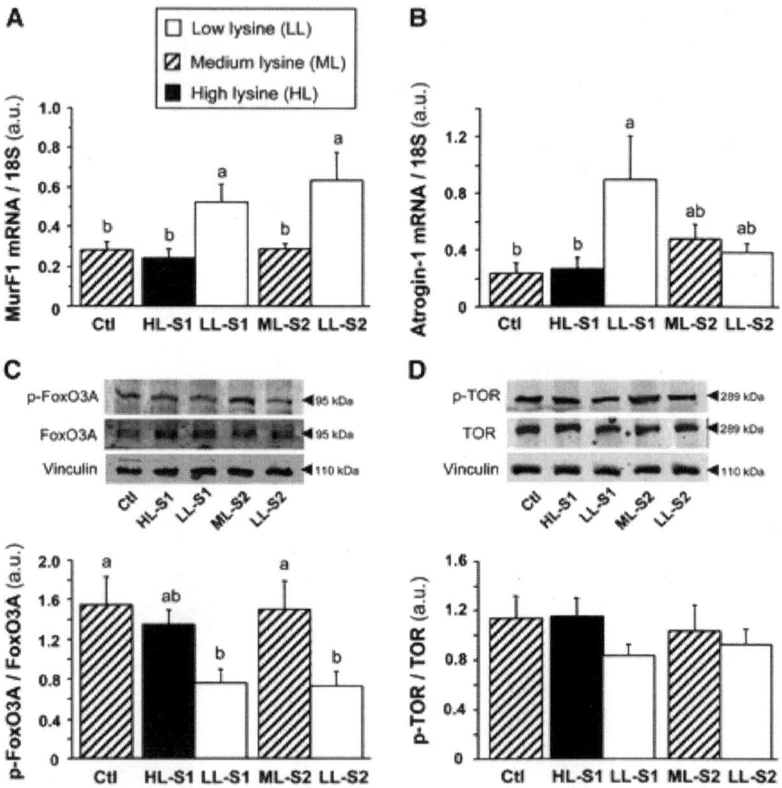

Figure: *: Effects of 48-h–cycle sequential feeding with diets varying in lysine content on the expression of MurF1* (A) *and atrogin-1* (B)*, and the phosphorylation of FoxO3A* (C) *and TOR* (D) *in breast muscle of 24-d-old chickens.* (A,B) *Analysis by real-time RT-PCR. mRNA levels were normalised using 18S rRNA; 18S was not significantly affected by dietary treatments.* (C,D) *Representative western blots of FoxO3A and TOR phosphorylation. Blots were quantified and the phospho-FoxO3A:FoxO3A ratio and phospho-TOR:TOR ratio were determined. Data are expressed as means ± SEM,* n *= 6. Means without a common letter differ,* $P < 0.05$*. Whatever the dietary treatment within a sequence, the indicated diet corresponds to the last diet in a cycle before tissue sampling.*

MurF1 mRNA levels were twice as high in the muscle of chickens fed the LL diet (S1 and S2) than in the HL-S1, ML-S2, and ML-Ctl groups (Atrogin-1 expression was twice as high in the LL-S1 group compared with the HL-S1 and ML-Ctl groups, but S2 did not differ between the LL and ML groups. To extend these findings, we investigated the regulation of MurF1 and Atrogin-1 in another original model of cyclical changes in diet composition (23). Interestingly, MurF1 and Atrogin-1 were overexpressed in the muscle of chickens transiently fed a low-protein diet compared with an isoenergetic high-protein diet.

FoxO3A and TOR Phosphorylation

MurF1 and atrogin-1 are regulated via mechanisms involving the protein kinase B/Forkhead box-O transcription factors (FoxO) and protein kinase B/mammalian target of rapamycin pathways (29–31). We first focused on the potential changes in FoxO3A phosphorylation induced by daily variations in dietary lysine content. The levels of phosphorylated FoxO3A were $<{}^{-}2$-fold lower in muscle of chickens fed the LL diet (S1 or S2) compared with the HL-S1, ML-S2, and ML-Ctl groups, suggesting that the overexpression of the E3 ubiquitin ligases (at least MurF1) in LL groups can be partly explained by significantly decreased FoxO3A phosphorylation. We then examined TOR phosphorylation. Despite lower phospho-TOR levels in the muscle of chickens fed the LL diet compared with the HL diet in sequence S1 (data not shown), we did not demonstrate a significant effect of daily lysine variation in that sequence when TOR phosphorylation on residue S2448 was normalised for total TOR content. Nevertheless, the correlation between FoxO3A phosphorylation and expression of the genes studied was significant for MurF1 ($r = {}^{-}0.362$; $P < 0.05$) but not for atrogin-1, suggesting quite different regulatory mechanisms for the 2 E3 ubiquitin ligases.

Chapter 7

Epidemiology of Newcastle Disease

This chapter is restricted to a consideration of Newcastle disease as it affects the birds that I refer to as village chickens, and that others regard as rural chickens or family chickens. I have been for many years project leader to programs funded by the Australian Centre for International Agricultural Research (ACIAR) and concerned with the development of Newcastle disease vaccines for use in village chickens. Of recent years there has been increasing interaction between the disparate organisations that have an interest in village chickens and the potential contribution of chickens to the alleviation of poverty amongst the rural poor. This group now includes FAO, IAEA, INFPD, World Bank, Danida and numerous NGOs. I welcome this opportunity to visit Denmark and meet with the Network for Poultry Production and Health in Developing Countries.

The view of ACIAR has been that little progress could be made with village poultry until Newcastle disease was controlled. Now that suitable vaccines are available it is possible to adopt a more holistic approach to the whole problem of rural chickens, and indeed of all rural poultry.

The Virus

The virus of Newcastle disease is classified within the genus *Paramyxovirus* of the family *Paramyxoviridae*. This immediately tells us certain immutable characteristics of the virus. The virus will have a genome of single stranded RNA. The inexact replication of the RNA will frequently produce variants with differences, often subtle differences, in phenotype from the parent particle. Unless there is suitable selection pressure, these variants will not prosper. We must be aware that the populations of Newcastle disease virus that spread in the field, or the populations that make up a vaccine stock, are not clonal. Selection pressure can alter the average behaviour of the population. Of particular interest to this discussion are the variations that can evolve in pathogenicity and in thermostability.

The infectious virus particle (the virion) will have a lipoprotein envelope that will be essential for infectivity. The proteins on the envelope will be specified by the viral genome. These will be antigenically important and they will contribute to the host specificity and the spectrum of pathogenicity of the virus. We will also be able to suggest other properties of Newcastle disease virus, by biological analogy with other paramyxoviruses. In particular, we will expect that Newcastle disease virus will be antigenically stable across its geographical range, and with time. Although variants will be detectable with monoclonal antibodies, or by sequence analysis, polyvalent antiserum will not easily distinguish between strains.

Newcastle disease viruses are usually cultivated in the cells lining the allantoic cavity of embryonated hen eggs. Some strains kill the embryos; others do not. The virus will also grow in cell cultures of avian origin, and in some mammalian cells. The replication of some strains of the virus is indicated by the destruction of the host cells, a process termed cytopathogenicity. Not all strains of Newcastle disease virus are cytopathic and detection of these strains in cultured cells can be difficult. All strains of Newcastle disease virus will agglutinate chicken red blood cells *in vitro* (and sometimes red blood cells from other species). The process is known as haemagglutination and is the basis of the common serological test, the haemagglutination-inhibition test, used to detect antibodies to this virus. Other serological tests are available.

Diseases as a Risk Factor in Relation to the Rural Poultry Model in Bangladesh

In addition to the question of breed quality of the chickens used in the SLDP, it has been underlined several times during evaluations of projects that the poor health status of the day-old chickens provided by the poultry model represents a major problem.

A number of people (avian pathologists associated the Department of Livestock Services' (DLS) poultry farms, directors of DLS farms, staff at Central Disease Investigation Laboratory (CDIL), Bangladesh Livestock Research Institute (BLRI), Bangladesh Agricultural

University (BAU), the NGOs BRAC and PROSIKA, consultants to the private poultry sector, chicken- and key rearers at village level) have all expressed that little improvement of the health status of chickens at most levels of the production pyramid of the poultry model has been obtained during the most recent years.

Important poultry diseases for any production system such as Newcastle disease, fowl typhoid, pullorum disease, fowl cholera, Gumboro, Marek's disease and fowl pox have all been diagnosed in Bangladesh and are suspected to be responsible for a high percentage of the significant mortality reported.

Other infections, which are known to influence production results negatively such as mycoplasma infections, infectious bronchitis, coryza and parasites, are probably common. A chicken rearer recently reported that the highest total mortality she had experienced in one of her batches of chickens (420) over a two-year period was approximately 36 % within the 8 week rearing period. Due to insufficient record and reporting systems no specific mortality rates from the government hatchery farms have been obtained but it is considered to be very high.

From time to time, the NGOs also experience high mortalities and disease outbreaks. In the private commercial sector where biosecurity has reached a higher level than at the government farms 15% mortality during rearing of the parent flocks is considered good. It is estimated that the mortality at most of the government hatchery farms at times can be significantly higher. Not all mortality is due to disease but such mortality rates clearly indicate that the disease situation represents a major problem in the breeding and multiplication system and therefore also to the rural poultry model. In this respect it can be mentioned that the diagnosis made often are based on a few uncertain criteria even at official laboratories. Newcastle disease, for instance, is often diagnosed only on the basis of bleedings of the alimentery tract. As a result of this other significant forms of Newcastle may be overlooked. Consequently, no information about the real situation of Newcastle in Bangladesh is available.

Biosecurity and Management Aspects

DLS Farms

Presently, the buildings and infrastructure at the government farms are not found suitable for breeding and support of breeding stock and the personnel at the farms are not well aware of "risky procedures" and good management practices.

"Open" or "semi-closed" poultry houses used for different purposes such as laying, rearing and brooding are often located within a very small radius on the farms resulting in a high risk for transmission of diseases between houses. The same houses are constructed with openings in the walls covered with wire, which are almost impossible to clean and disinfect. The open houses are also very vulnerable to diseases, which are known to have a reservoir in the wild life. These include influenza, Newcastle disease, salmonella infections and fowl cholera. None of the houses have shower in facilities for the personnel.

Eggs and chickens of various ages are exchanged quite intensively between the different DLS farms and represent a serious risk of spreading of diseases. The reason for this exchange is the lack of an overall production and multiplication plan.

The staff at the farm goes in and out of the different houses without any precautions and no fixed procedures have been implemented. Simple precautions such as the use of foot-dips or change of clothes etc. at the entrance to all houses are not taken.

Vaccines (primarily Gumboro) are still not always available. When the farms run out of vaccine stock the application procedure for supply of new batches of vaccines may take a long time with a resulting temporary lack of vaccine. Another problem is that efficacy testing has not been carried out on vaccines produced by DLS and no consistent vaccination policy has been made. Vaccination against Marek's disease is only performed in case of acute problems. But the diagnosis of Marek is based on insufficient pathological criteria, which may result in wrong conclusions. Storage and use of vaccines are not performed strictly according to the manufactures' recommendations.

No serological or bacterial routine monitoring of the flocks are performed before the start of laying or during the laying period. Routine post mortems of dead birds are not performed.

In addition, the layout of the hatcheries does not always comply to sound biosecurity standards, and the staff at the hatcheries are not aware of common hygienic precautions needed to be taken when handling hatching eggs, including proper disinfection and storage. The lack of disinfection of hatching eggs under the conditions existing in Bangladesh should be regarded as unacceptable.

NGO Poultry Farms

BRAC: The most recent farms build by BRAC are modern parent stock farms constructed according to high biosecurity standards. That is, closed, fully ventilated houses where the rearing unit is located at a reasonable distance from the production houses. Shower-in facilities are used. The associated hatcheries are also distant from the poultry units and have their own entrance shower-in facilities etc. The layout of the hatcheries is according to highest standards and planning has been done in collaboration with the manufacturer of the hatchers.

However, also semi closed production systems with a resulting minor level of biosecurity exist within the production system of BRAC (two farms). Farm management has a high priority according to BRAC and internal courses are frequent. The farm managers meet at one of the BRAC centres once a month to coordinate working routines and standards. Staff is sent to the manufacturer of the setters and hatchers for training in hatchery management. Due to the lack of qualified local veterinarians within poultry production, BRAC has employed a veterinarian from abroad. He is visiting the farms, doing post mortems and trains the staff etc.

BRAC is interested in setting up a monitoring and control system, but does not have the possibility with the existing laboratory facilities. According to BRAC there is an urgent need for diagnostic laboratory capacity in Bangladesh and they are constructing their own laboratory, which is expected to be operational late summer 1999. In the meantime serology (mainly ELISA) is performed at a laboratory run by a private poultry company.

PROSIKA: Currently PROSIKA has two parent-stock farms in operation. One modern closed housing farm and another semi-closed farm. Both farms contain a mixed population of commercial hybrid parent-stock for broiler and layer production.

In 1999 two more closed housing system farms are expected to be fully operational. Both these new farms will be parent-stock farms with a mixed population of commercial hybrid parent-stock for broiler and layer production. Training of farm managers is performed abroad by the breeding company supplying the parent-stock. Hatchery management training has to some extend been done in India. The overall impression of farm management/level of biosecurity etc. is that PROSIKA do take precaution but they to could improve on these issues.

This organisation has also expressed the need for extended diagnostic services.

Laboratory Back-up Facilities

The nucleus of "Veterinary Investigation Service in Bangladesh" is CDIL located in Dhaka, but also eight "Field Disease Investigation Laboratories" (FDIL´s) located at Mymensingh, Manikganj, Joypurhat,

Barisal, Gaibanhda, Serajgonj, Sylhet and Feni are part of the system. The main tasks of the CDIL are to examine specimens received from the field veterinarians, farmers, government and private livestock and poultry farms or any other source and to render diagnosis along with appropriate line of treatment and suitable control measures. The CDIL receives specimens mostly from rural and urban areas in and around Dhaka. It also receives specimens from the FDIL´s.

The technical knowledge and procedures required to perform a simple but still beneficial disease monitoring and control programme for the breeding flocks appear to be present at CDIL (mainly monitoring the salmonella-, mycoplasma-, Newcastle disease virus status of the breeding flocks). However, it is impossible for the present number of scientific staff at CDIL to run the number of test required in a monitoring and control programme in parallel with their diagnostic duties. The other laboratories will have major problems in implementing the procedures needed. In general, there is a significant lack of chemicals, reagents, media etc. at these laboratories just to be able to solve the present diagnostic tasks.

The support, which would be needed from these laboratories in case of the introduction of a monitoring and control system of the breeding flocks, would further increase the demand for media and chemicals. More detailed diagnosis and subsequent control measures cannot be obtained from any of these laboratories at present. Newer diagnostic tools such as commercial available ELISA tests are not available. As a consequence of this, one private poultry breeding company has started a serological laboratory where several ELISA tests are performed. BRAC is using this laboratory from time to time.

Future Actions Recommended to be Taken

Within the DLS production system, a new production setup is needed in order to be able to control the production as well as the disease situation. It is recommended that the DLS farm located at Savar operates as a central grandparent farm and the other DLS hatchery farms as parent-stock units. Attention should be given to the fact that housing of birds of different ages on the same site increase the risk of persistence and cross-infection and reduces or prevents the opportunity of achieving a total depopulation and proper cleaning of the site. On a multi-age farm it is also more difficult to maintain effective live vaccination programmes.

It is recognised that the DLS hatchery farms probably cannot be run as all-in, all-out sites, but due to the high disease pressure which is suspected at these government farms it is strongly recommended that the sites as a minimum are depopulated and cleaned and disinfected prior to the arrival of stock within a new production structure. This could be done for one farm at a time. As soon as possible after a house has been depopulated, dry cleaning followed by wet cleaning and disinfection, and sanitisation of the drinking- and feeding systems according to a written protocol should be performed.

Traffic to and on farm and general hygiene, handling of equipment, rodent and pest control, hatchery management and hygiene, feed storage and vaccination are all important aspects, which need to be addressed by DLS, in particular.

The NGOs and DLS should work together in setting up health monitoring and disease control programmes of breeding flocks which initially should seek to eradicate pullorum disease, fowl typhoid, mycoplasma infections, Newcastle disease, Gumboro and leucosis from the top of the production pyramid.

Training of personnel at all levels of the production (including hatchery operation) is necessary to be able to achieve significant improvements. Upgrading of laboratory back-up facilities must have a high priority with special emphasis on increasing the number of scientific staff. Proper use of vaccines and consequent use of vaccination programmes is like-wise needed and require the availability of the relevant and efficient vaccines. Research in the significance, prevalence, epidemiology etc. of diseases should be initiated as recommended in the "Applied research and study component poultry development" (Darudec, 1998).

In the following, a more detailed outline of important aspects to be addressed will be given.

Traffic and Hygiene

Traffic onto the farm is a major source of disease introduction, and humans are the single most important factor in the form of e.g. truck drivers, sales people and service crew who routinely go from farm to farm and flock to flock in a short period of time. The following can control human traffic at relevant farms:

- Keep all nonessential personnel off the farm.
- Establish a logbook and record non-farm staff that enters the farm
- It should be made possible to distinguish "dirty" and "clean" areas on the farm

- Confine employee parking to the dirty side of the farm
- Lock all doors and use gates to prevent unwanted vehicle traffic.
- Appropriate procedures should be taken when crossing from dirty to clean area, such as showering or change clothing and footwear.

In addition, it is strongly recommended that descriptions of the following routine operations at the farms are available for the staff and that they are understood:

- Egg collection
- Collection of spent hens
- Arrival of new stock

Equipment

Equipment used in handling, managing and moving birds, feed, eggs or manure is capable of mechanically transmitting diseases from one location to another. Movement of equipment from one farm to another should consequently be avoided. Trucks, particularly load-out trucks and associated equipment, can be heavily contaminated and should if possible be avoided on the farm. If they have to enter they should be carefully washed and disinfected before entering the farm. Rodents and other pests. Wild birds and other animals are significant sources of poultry diseases such as avian influenza, Newcastle disease and fowl cholera. Rodents and different beetles may be important sources of salmonella, *E.coli*, infectious bursal disease (Gumboro) and fowl cholera. Flies can transmit Marek's disease and intestinal worms. The following measures should be used to reduce these pests on the farm.

- Implement rodent and insect control programmes
- Screen and eliminate nearby nesting areas for wild birds
- Keep the vegetation round the houses down
- Keep litter as dry as possible to prevent flies from laying
- Eliminate all access points into the houses (this require of course closed housing system)

Hatchery Management Considerations to be Made

Location. Should be located at a distance from the poultry farm. (This is not always the case at the DLS hatchery farms).

Layout. Careful planning of the working areas of the hatchery can greatly assist in allowing hygienic work practices by separating "clean" and "dirty" areas.

Ventilation. Should be operated to support separation of clean and dirty areas.

Waste disposal. Proper arrangements for the disposal of hatchery waste will reduce the risk of contaminating nearby poultry units.

Site security. As with farms, proper security and control of visitors is essential cleaning and disinfection procedures of site as well as equipment.

Hand hygiene :of staff is important throughout the various stages of handling hatching eggs

Written procedures: for most working routines including collection of eggs at the farms.

Hatching & Hatchery Hygiene

The egg laid, is vulnerable to damage and bacteria and should be collected quickly. In that respect the number of floor eggs should be reduced by proper actions such as to ensure a sufficient number of nests placed optimally in the houses. It has been observed that this could be improved at the government hatchery farms. Production planning is important to be able to produce DOC in right numbers at the right time. Better timing and planning could be obtained in Bangladesh if better analysis and arrangements were made in collaboration between the NGO´s and the government farms. God hatching eggs must:

- be fresh (stored less than 7 days)
- be stored under optimal conditions (15-20^{0}C, 75% relative humidity)
- be properly disinfected as soon as possible following collection. This is very important under Bangladesh conditions with the high temperatures and high humidity experienced. Growth of fungi has been reported to be a significant problem at the hatcheries causing poor hatchability, contamination of newly hatched chicks with subsequent high mortality.
- be free of transmissible pathogens (such as mycoplasmas, salmonella, leucosis virus etc.)
- have a uniform, sufficiently high level of maternal antibodies as ensured by proper vaccination.

Feed

Besides quality control of feedstuff, measures should be taken to prevent contamination of feed with pathogens. Food stores should be bird and rodent proofed to prevent contamination with e.g. Newcastle disease virus from wild birds and salmonella. Stores should be cleaned between flocks.

Monitoring of the Flock's Health Status

The following should be evaluated and recorded daily:

- Water and feed intake
- Mortality and morbidity and suspected causes
- Bird behaviour and appearance (beak, eyes, feathers, vent, legs and droppings)
- Presence of external parasites.

Considering the poultry health situation in Bangladesh it is strongly recommended that a disease control programme be implemented in a new production system. It should include the following aspects but must not be taken as a complete programme. The setting up of such a programme would require further analysis

Pullorum Disease and Fowl Typhoid

As these diseases are transmitted vertically to the progeny, it must be sought to eliminate these infections in top of the production pyramid. Consequently, repeatedly blood testing of breeding flocks and culling of reactors must be performed. All birds at the breeding farms should be tested twice between 16 weeks and point of lay and subjected to consecutive clear tests 1 month apart.

An annual clear test of 60 birds/flock should be initiated. Screening should also be performed at the hatchery farms. The need requires further analysis. The test used can be the rapid plate agglutination (RPA) test already used by CDIL.

Mycoplasmas

The microorganisms are also transmitted through the egg and elimination is the goal. Before the GP and P flocks come into lay in should be tested by rapid serum agglutination test (RSA) for *Mycoplasma gallisepticum* and *synoviae* infection. This test is already used by CDIL. A programme of eradication should be established and leucosis could be included.

Newcastle Disease

Proper vaccination programmes should be established based upon local experiences. Further analysis of the situation is required before final recommendations can be made. Such analysis may actually save vaccine. Today, some of the farms vaccinate once a month. This could be avoided by clarification of the problems.

Gumboro Disease

It would be most helpful in the understanding of the Gumboro problem to use the ELISA test in the breeding flocks to obtain information on the size of the problem. A proper vaccination programme should be implemented based upon management used and virulence of strains prevalent.

Vaccination

It should be kept in mind that the first step in a good vaccination programme is proper management including biosecurity. It further depends on:

- Storage condition of the vaccine, including temperature, time and time lag between preparation and utilisation
- Concentration of the vaccine
- Application method
- Age of the birds
- Immune status of the flock
- Disease pressure

A vaccination programme for the breeders should initially include:

- Marek
- Newcastle
- Gumboro
- Fowl pox
- (Cholera).

How to start this effectively and the extent of it depends upon:

- The establishment of a disease surveillance programme to make proper disease identification
- Detailed management and biosecurity analysis
- Logistic investigations enabling risk assessment.

It is recognised that this cannot be implemented all at once, but it should be aimed for in the future just as the use of other vaccines. In addition to the periodical lack of vaccine supply there does not seem to be efficacy control with vaccines produced by DLS including Newcastle vaccine. It is strongly recommended to set up a control scheme for these self-produced vaccines.

Another aspect to realise is that, under field conditions vaccinations alone is insufficient to bring about effective control of Newcastle disease and other diseases and must therefore be accompanied by good hygiene.

In poorly managed, over-crowded, badly ventilated conditions with inevitable underlying bacterial infections even the mildest vaccine strains may produce severe disease.

A general Review on some Important Diseases in Free Range Chickens

Poultry production has undergone rapid changes during the past decades due to the introduction of modern intensive production methods, new breeds and improved biosecurity and preventive health measures. Moreover, the management methods place high demands on proper health, hygiene and require only a small, but very skilled labour force.

In developing countries, however, adoption of this type of production has been limited due to the high inputs as listed above. The progress in industrial poultry production methods has thus had little effect on subsistence poultry production in the rural and periurban areas in developing countries. In these countries access to poultry meat and eggs depends on village-level poultry production.

Although poultry production is considered as secondary to other agricultural production systems it has an important role in supplying villagers with additional income and high quality protein. This system provides valuable protein through a low input system, now representing 30% or more of the protein consumed (FAO, 1998). Almost all families in developing countries keep a chicken flock with an average size of about 10 adult chickens, varying from 5 to 50 animals.

The majority of these animals are kept in free range scavenging systems, where the birds scavenge around the house during daytime. Primitive housing of the birds during the night, however, is often seen. Supplementary feed consists mainly of household wastes, insects, larvae and seeds. Mortalities observed are in the range of 80 - 90% within the first year after hatching. For the same reasons the owners never include chicks when they refer to the flock size. The mortality is believed to be caused by mismanagement, lack of supplementary feeding, predators and diseases.

Little research has been published on rural poultry health, despite the fact that up to 80% of the poultry population in Africa and Asia is kept by the households as free range chickens.

Although solid data have not been published, Newcastle Disease (ND) is regarded as the principle factor limiting rural poultry production in all African and Asian countries. ND may kill up to 80% of household poultry in Africa, but is not expected to account for the high early mortality rate according to the authors.

In addition, detailed epidemiology of the disease in the village situation is largely unknown (Yongolo, 1997).

Furthermore, recent studies have shown that other diseases are present in scavenging poultry communities.

Since most of our knowledge relies on seroprevalence studies, solid longitudinal studies on causes of mortality are strongly needed to improve our knowledge of the prevalence and significance of the individual diseases under village conditions.

The following data therefore mainly reflect experience obtained under backyard conditions in developed countries.

Diseases

According to Jordan et al. (1996) and Calnek et al. (1997) poultry diseases can be divided into five groups, namely bacterial, viral, fungal, parasitic and nutritional diseases.

Only the diseases of expected importance under village conditions, e.g. those causing high mortality rates in chickens are mentioned in these tables.

As seen a wide variety of diseases are expected to occur under village conditions. Some of these diseases are age specific, whereas others are encountered in all age groups.

Table: *Important bacterial diseases in free range poultry and the age group where the disease is most often observed*

Disease	***Age group***
Escherichia coli	All ages, but mainly chicks
Salmonella spp.	All ages, but mainly chicks
Salmonella pullorum	Chicks < 3 weeks
Salmonella gallinarum	Growers, adults
Pasteurella multocida	Growers, adults
Haemophilus paragallinarum (Coryza)	Growers, adults
Clostridium perfringens	All ages, but mainly growers
Mycobacterium avium	Adults
Mycoplasma gallisepticum	All ages
Mycoplasma synoviae	All ages

Table: *Important virus diseases in free range poultry and the age group where the disease is most often observed.*

Disease	***Age group***
*Marek's disease	> 6 weeks
*Leucosis	Adults
Newcastle disease	Mainly growers and adults
Fowl Pox	All ages
Infectious Laryngotracheitis	Growers, adults
*Infectious Bursal Disease "Gumboro	< 8 weeks
*immunosuppressive disease	

Table: *Important fungal diseases in free range poultry and the age group where the disease is most often observed.*

Disease	***Age group***
Aspergilloses	Chicks
Mycotoxicoses	All ages

Table: *Important parasitic diseases in free range poultry and the age group where the disease is most often observed.*

Disease	***Age group***
Coccidiosis	Chicks, growers, (adults)
Histomoniasis	1-3 months
Nematodes	All ages
Haemoparasites	Chicks, growers
Ectoparasites	Chicks, growers

Table: *Important nutritional diseases in free range poultry and the age group where the disease is most often observed*

Disease	***Age group***
Vitamin A, D & E	Chicks, growers
Other vitamins, minerals and amino acids	Chicks, growers

Discussion

Approximately 80% of the world poultry population is kept as free range poultry. The free range poultry production system has also been designated as the "low input - low output" system (Pandey, 1992). Mortality in this system is in the range of 80 - 90% within the first year after hatching and is believed to be caused by mismanagement, lack of fresh water and supplementary feed, predators and diseases. Of these, diseases are believed to be the main limiting factor to the production of indigenous chickens (Aini, 1990). Among causes of early mortality nutritional diseases might be expected to dominate due to shortage of supplementary feed before and after hatch.

In addition, the quality of hatching eggs might be questioned under the climatic conditions present in these countries. Lack of vitamins and protein weaken the chicks and make them vulnerable to other diseases and predators. Diseases are also easily contracted under free range conditions due to scavenging habits. With an unconfined type of management, disease control is very difficult to carry out and is therefore rarely practised by the owners.

As mentioned earlier, Newcastle Disease is believed to be the most important disease in free range systems. During out-breaks of the disease up to 80% of the population may die. This, however, is dependant on different factors including the virulence of the strain causing the out-break (Alexander, 1997). A recent study in Nicaragua has, however, shown that in ND-immunised birds mortality is still high. The majority of the mortality is found in chicks up to 3-4 month of age. In this group up to 52.5% of the animals died due to other causes than ND. Similar studies in Mali by Wilson et al. (1987) have shown that chick mortality is in the range of 60% within the first 3 months after hatching.

A study in Morocco has revealed that up to 58% of the village chickens had antibodies against Salmonella gallinarum and S. pullorum. Similar findings were reported by (Adesiyun et al. 1984) from Nigeria. Chryosostome and his co-workers (1995) reported that 10% of the village chickens had antibodies against S. pullorum and that 62% had antibodies against Mycoplasma gallisepticum. Furthermore, 65% of the animals had antibodies against ND. In Mauritania, Bell et al. (1990) found that 17.5% of the birds had antibodies against S. pullorum and that up to 46.2% of the birds had antibodies against Gumboro disease. In the same animals 7.5% had antibodies against ND. In Tanzania, Permin et al. (1999) examined 600 live chickens and found the presence of a range of diseases. All animals were parasitised with one or more (up to 14 species) species of endoparasites.

In total 29 different species were detected in the study. Furthermore, 65.7% of the animals were parasitised with Cnemidocoptes mutans, Dermanyssus gallinae and/or Echidnophaga gallinacea. The animals were also infected with a range of haemoparasites, the most common being Plasmodium juxtanucleare and Aegyptinella spp.

Antibodies against Newcastle disease was seen in 7.3%, against Salmonella enteriditis in 2.0%, against Salmonella gallinarum/pullorum in 52.7%, against Infectious Laryngotracheitis in 58.3% and against

Gumboro disease in 42.3%. Similar studies have to the knowledge of the authors not been carried out in Asian countries. The significance of all these diseases, however, remains to be investigated. In addition, it should be noted here that a general trend for these studies is that they have only looked for antibodies against selected diseases.

Long-term cohort studies, examining the causes of death, have to date not been carried out in the free range production systems. Important knowledge on the proportion of the individual disease of the overall mortality is thus not known. It is consequently postulated that other diseases than ND are present in free range poultry production systems and that a successful development of this production system is only achieved when the exact causes of death is known. Since publications on disease prevalence based upon post mortem examinations are not expected to be accepted by existing international peer-reviewed journals the W.P.S.A. should be addressed to establish an international journal dealing specifically with problems relating to scavenging and family poultry. Establishing such a journal would ensure a rapid implementation of research results obtained.

"This is the first ever such atlas to be produced for Indian farmers and poultry professionals. Its objective is to provide a handy tool to farmers for on-the spot diagnosis of diseases, under field conditions.

Although a large number of laboratory tests, some sophisticated like ELISA and PCR, have been developed for the diagnosis of poultry diseases, they are either not available or are beyond the reach of most farmers. Moreover, by the time birds reach laboratory, they are mostly decomposed and unfit for postmortem examination and disease diagnosis. The delay in diagnosis allows the disease to inflict mortality and ruin farmer's economy. The key to disease control then is on-the spot diagnosis. It is with this objective that the colour atlas has been produced-to make readily available to poultry farmers a handy tool for immediate and accurate disease diagnosis under field conditions.

A large number of poultry diseases, including the more deadly like Ranikhet and Gumboro, leave their footprints in the dead bird. From these characteristic postmortem findings, it is possible to arrive at a correct diagnosis. The atlas provides 150 coloured photographs and covers a wide spectrum of poultry diseases. For important diseases, several photographs are given so that at no stage their diagnosis is missed. Also, wherever considered appropriate, pictures of live birds showing symptoms of specific diseases are given. In addition, each disease is accompanied by a brief description that highlights its salient features and also records author's field observations.

As this book is intended exclusively for disease diagnosis, it should be read alongside author's another book - "Poultry diseases - A guide for farmers and poultry professionals (2004) by J.L. Vegad" - to obtain information on other aspects of poultry diseases, namely, cause, transmission, symptoms, diagnosis, treatment, and control.

Although the book is written to meet requirements of poultry farmers, it will be equally useful to poultry consultants, diagnostic laboratories, and will acquaint the veterinary students with poultry diseases under field conditions."

The world's industry makes a significant contribution to the supply of dietary animal protein in both industrial and developing countries.

The structure of India's poultry industry varies from region to region. While Independent and relatively small-scale producers account for the bulk of production, integrated large scale producers do account for a growing share of output in some regions. Integrators include large regional firms that incorporate all aspects of production, including the raising of grandparent and parent flocks, rearing DOCs, contracting feed, providing veterinary services and whole selling.

Although the world's poultry industry has recorded unprecedented expansion during the past two decades, constraints associated with overproduction, environmental pollution and restraints to international trade are influencing the rate of expansion in various regions.

The major factors influencing the poultry industry are climate, availability and cost of land, capital and labour, proximity to markets, competitive efficiency of integrators, consumer satisfaction through safety, quality and flock welfare, environmental acceptability of large operations and successful implementation of international trade and above all, the price of feed ingredients determine the profitability of poultry production.

1. Guide to Commercial Poultry Production.
2. Improving backyard Poultry-keeping.
3. Knowledge-level among Poultry entrepreneurs on scientific layer Farming.
4. Issues and Economics of Poultry Production.
5. Fowl Play: The poultry industry's central Role in the bird flu crisis.
6. The use of Vaccination in Poultry Production.
7. Use of Herbal Plants in Poultry Health Management in the small-scale Commercial Farming.
8. Profitable Poultry: raising birds on pasture.

9. Biochemical and histopathological analysis of Aflatoxicosis in growing Hens Fed.
10. Patterns of change in Poultry Production Industry.
11. Exposure to dust and bioaerosols in Poultry Farming.
12. Guidelines for the Establishment and Operation of Poultry Farms.
13. Poultry Production Management and Bio-security Measures.
14. Environmental controlled Poultry Farm.
15. Poultry Processing opportunities.
16. Pastured Poultry Production: An Evaluation of its sustainability.
17. Strategy of household Poultry Development in coastal regions.
18. Small Commercial and family Poultry Production (A Case Study).
19. Organic Poultry Products are they the real deal.

Poultry farming is the practice of raising poultry, chickens, turkeys, ducks, and geese, as a subcategory of animal husbandry, for the purpose of farming meat or eggs for food. More than 50 billion chickens are reared annualy as a source of food, for both their meat and their eggs. Chickens farmed for meat are called broilers, whilst those farmed for eggs are called egg-laying hens. In total, for eggs are called egg-laying hens. In total, the UK alone consumers over 29 million eggs per day. Some lens can produce over 300 eggs a year. Chickens will naturally live for 6 or more years. After 12 month, the hen's productivity will start to decline. This is when most commercial laying hens are slaughtered.

Economic Raising of Poultry (Broiler, Layer, Ducks, Quail and Turkey)

Poultry Layer Farming

Poultry egg and meat are important sources of high quality proteins, minerals and vitamins to balance the human diet. Specially developed breeds of egg type chicken are now available with traits of quick growth and high feed conversion efficiency. Depending on the farm-size, layer (for eggs) farming can be main source of family income or can provide income and gainful employment to farmers throughout the year. Poultry manure has high fertilizer value and can be used for increasing yield of all crops

Scope for Layer Farming and its National Importance

Poultry industry which provides cheap source of animal protein has taken a quantum leap in the last three decades evolving from a near backyard practice to a venture of industrial promotion. Poultry is one of the fastest growing segments of the agricultural sector in India today. While the production of agricultural crops has been rising at a rate of 1.5 to 2 percent per annum that of eggs has been rising at a rate of 8 percent per annum. India is on the world map as one of the top five egg producing countries with 55.6 billion eggs produced during 2008 (FAO).

The poultry sector in India has undergone a paradigm shift in structure and operation. This transformation has involved sizable investments in breeding, hatching, rearing and processing. Farmers in India have moved from rearing nondescript birds to rearing hybrids which ensures faster growth, good liveability, excellent feed conversion and high profits to the rearers. High quality chicks, equipment, vaccines and medicines are available. Technically and professionally competent guidance is available to the farmers. The management practices have improved and disease and mortality incidences are reduced to a great extent. The industry has grown largely due to the initiative of private enterprise, minimal government intervention, considerable indigenous poultry genetic capabilities and adequate support from the complementary veterinary health, poultry feed, poultry equipment and poultry processing sectors. The industry has created direct and indirect employment for 3 million people.

Financial Assistance Available from Banks/NABARD

Loan from banks with refinance facility from NABARD is available for starting poultry farming. For poultry farming schemes with very large outlays, detailed project reports will have to be prepared. Banks provide financial assistance for the following purposes :

a. For construction of brooder/grower and layer sheds, feed store, quarters etc.

b. For purchase of poultry equipment such as feeders, waterers, brooders etc.

c. For creating infrastructure items for supply of electricity, feed, water etc.

d. For purchase of day old chicks or ready to lay pullets.

e. For meeting working capital requirement in respect of feed, medicines and veterinary aid etc. for the first 5 to 6 months (i.e. till the stage of income generation).

The Cost of land is not considered for loan.

Scheme Formulation for Bank Loan

A scheme can be prepared by the beneficiary after consulting local technical persons of State Animal Husbandry/Veterinary department, Poultry Corporation or private commercial hatcheries. If possible, they should also visit the progressive layer farms in the area and discuss the profitability of farming. A good practical training and experience on a layer farm will be highly desirable, before starting a farm. Regular and constant demand for eggs and nearness of the farm to the market should be ensured. The scheme should include information on land, water and electricity facility, marketing aspects, training facilities and experience of entrepreneurs and the type of assistance available from State government, poultry corporation, local hatcheries. It should also include data on proposed capacity of the farm, total cost of the project, margin money to be provided by the beneficiary, requirement of bank loan, estimated annual expenditure, income and profit and the period for repayment of loan and interest. A format developed for project report preparation for a commercial layer farm is given in Annexure -I.

Appraisal of Project

After the scheme is submitted to the bank it is examined for technical feasibility and economic viability.

Technical Feasibility : This would briefly include :-

a. Suitability of climate and potentiality of the area

b. Availability of inputs such as chicks, feed, medicines etc.

c. Technical norms

d. Infrastructure available for veterinary aid, marketing, training and experience of the beneficiary.

Financial Viability : This would briefly include :-

a. Unit cost and loan requirement.

b. Input cost for chicks, feed, veterinary aid, labour and other overheads.

c. Output cost i.e. sale of eggs, culled birds, for meat, manure, empty gunny bags etc.
d. Income-expenditure statement and annual gross surplus.
e. Cash flow analysis.
f. Repayment schedule i.e. repayment of principal loan amount and interest.

Other documents such as loan application form, security aspects, margin money requirement etc. are also examined. A field visit to scheme area is undertaken for conducting technoeconomic feasibility study for appraisal of the scheme.

Sanction of Bank Loan and its Disbursement

After ensuring technical feasibility and financial viability, the scheme is sanctioned by the bank. The loan is disbursed in kind in 2 or 3 stages against the creation of specific assets such as construction of sheds, purchase of equipment and machinery, recurring cost during growing period on purchase of chicks, feed, medicines and vaccines, electricity and water, labour expenses etc. for first cycle. Constant follow up and supervision of the scheme is done by the bank.

Lending Terms - general

Financial Outlay

Financial outlay of the project depends on the local conditions, unit size and the components included in the project. Prevailing market prices may be considered to arrive at the outlay.

Margin Money: Margin depends on the category of the borrowers and may range from 5 to 25% of the total outlay.

Interest Rate for ultimate borrower : Banks are free to decide the rates of interest within the overall guidelines. However, for working out the financial viability and bankability of the model projects we have assumed the rate of interest as 12 % p.a.

Security

Security will be as per NABARD/RBI guidelines issued from time to time.

Repayment Period of Loan

Repayment period depends upon the gross surplus from the project. The loan will be repaid in suitable monthly/quarterly instalments usually within a period of seven to nine years with first year as grace period.

Insurance

The birds and other assets (poultry sheds, equipments) may be insured. Wherever necessary, risk/mortality fund may be considered in lieu of insurance.

List of Medicine, Vaccine and Nutrient Manufacturers

Factors in Planning A Poultry Operation

Independent or Contract Production :

1. Independent egg producers have the total responsibility for all planning, producing and marketing of eggs. They must provide houses, equipment, birds, feed, all supplies and management. Their risks are greater but potential profit or loss is also greater. There are no independent broiler producers.
2. Contract poultry producers (broilers and eggs) provide land, labour, houses, equipment, taxes, utilities, and insurance. The contractor furnishes birds, feed, medication, supervision, and markets the product. The contractor picks up eggs or broilers at the farm and the producer is paid a base price per dozen or per pound plus an incentive bonus as stated in the contract.

In planning for a poultry operation, a determination should be made of all contractors operating within the area and the type of contract available.

Location within Florida

Egg farms are located in counties all over the state; however, the major egg producing areas are the Tampa Bay area, all across Central Florida, northward to Nassau County, and scattered independent producers located across the northern part of the state. Independent egg farms could be located in any county, however, make sure a market is available and zoning laws permit poultry farming. It would be well, however, to consider locating new farms well away from populated communities. All commercial broiler producers are located in North Florida (counties north of Gainesville). It would not be feasible to establish a broiler farm in other counties, since essentially all broilers are produced under contract and, at present, contractors only operate in North Florida.

Investment

With building and equipment costs continuing to rise, it is difficult to set an accurate investment cost necessary to build and equip poultry houses. Today's investment for buildings and equipment alone are running approximately \$5.00 - \$7.00 per bird for layers, depending on the degree of mechanisation. Broiler houses and equipment will run approximately $3.50 per square foot. Broilers are usually housed at about 0.8 square feet per bird. Independent egg farms may be any size. New egg farms usually have 50,000 layers per farm. Broiler farms have 25,000 - 90,000 bird capacity per farm.

Market

Since Florida is now an egg sufficient state, farms must produce eggs for a specific market. They can no longer produce eggs and then hope to sell them for a profit if they have to take a wholesale price for all eggs. In planning for an egg operation, the most important consideration is to have a specific market or a firm contract from a producer. One of the biggest advantages which a farmer has in producing eggs under contract is the freedom from concerns of marketing.

Labour

Table: *Suggested Number of Layers per Man on Independent Farms Under Varying Conditions Marketing Type***

*Producer Type**	*A*	*B*	*C*	*D*	*E*
Type 1	13,000-17,000	9,000-11,000	7,000-9,000	6,000-8,000	5,000-7,000
Type 2	8,000-10,000	7,000-8,000	5,000-6,000	4,000-5,000	3,500-4,000
Type 3	7,000-8,000	6,000-7,000	4,000-6,000	3,500-4,000	3,000-3,500
Type 4	7,000-8,000	6,000-7,000	5,000-6,000	4,000-5,000	3,500-4,000
Type 5	6,000-7,000	5,000-6,000	4,000-5,000	3,500-4,000	3,000-3,500
Type 6	5,000-6,000	4,000-5,000	3,500-4,000	3,000-3,500	2,500-3,000

**Marketing Types as Follows:

A = Selling at wholesale with no grading involved

B = Selling at wholesale with grading being done at the farm

C = Semi-retailing (case lots) with no candling or cartoning

D = Semi-retailing (case lots) when candling and cartoning

E = Retailing door to door

*Producer Types as Follows:

1 = Producer who buys started pullets and whose operation is fully automatic - with pit cleaners, egg gathering belts, etc.

2 = Producer who buys started pullets and has automatic feeders

3 = Producer who buys started pullets

4 = Producer who raises his own pullets and is fully automatic with pit cleaners, egg gathering belts, etc.

5 = Producer who raises his own pullets and has automatic feeders

6 = Producer who raises his own pullets

The most successful poultry operations are those in which the owner and the owner's family are actively involved in managing and working on the farm. Since poultry farming requires some work every day, the manager is responsible for labour seven days per week and most understand the need for this constant attention to details. The most successful poultry farmers enjoy this type of work.

Commercial Egg Production

Breeds and Strains Selection

1. Contract production - The contract egg producer usually has no voice in determining which strain of bird will be placed on the farm. The contractor will select the best bird for egg market.
2. Independent production - There are several top strains of egg production layers available. The producer should pick the strain to fit the market demand, such as egg size, egg shell colour, shell quality and feed efficiency. Chicks should be obtained from reputable hatcheries. By studying the Official Directory of Florida Breeding Flocks and Hatcheries the producer can find a source of quality chicks.

Housing and Equipment

Essentially all commercial layers in Florida are housed in cages. If an egg farm is going to produce eggs under contract, the producer will be able to get construction details from the contractor and often must build and equip the house following the contractor's directions. Since many contractor-owned complexes are built with closed sided, evaporative cooled houses, contract producers may be required to provide similar facilities.

Poultry House Essentials :

1. Locate on well drained land. High, sandy, black jack oak ridges make ideal locations.
2. Build on an elevated grade with good drainage between houses.
3. Consider acreage and location of buildings for future expansion.
4. Open houses should be built with ridge ventilation which can be opened and closed.
5. If houses are equipped with side curtains, the curtains should be adjustable.
6. Adequate house eaves (24-36 inches) to keep out blowing rain and direct sun are important.
7. Consider ease of manure clean out.

8. Make sure adequate land is available for manure disposal (approximately 1 ACRE/1000 birds).

Feeds and Feeding :

1. Feed and feeding instructions are provided by the contractors for farmers producing eggs under contract.
2. The independent producer should look at economics and buy high quality feed from a company offering good service as well as quality feed. Feeds are composed of important groups of ingredients called nutrients. A high quality feed will contain a proper balance of nutrients balanced for the purpose for which it is to be fed.
3. Most independent producers should start out buying complete feed. As he grows in size and gains experience, he may want to consider mixing his own feed. Investigate the economics, time and management necessary to mix a quality feed and maintain quality control. Details in formulation and feed mixing can be obtained from the Extension Poultrymen.
4. A feeding program should be developed to maintain quality. Check feed tanks for old, caked feed. Empty and clean tanks occasionally. Do not allow food to get wet.

Flock Health :

1. The contractor service person is trained to spot health problems before they develop into major problems and advise the producer.
2. The independent egg producer can best maintain healthy flocks by:
   - (a) Using the all-in-all-out method of management.
   - (b) Practising good management.
   - (c) Keeping his farm and houses off limits to visitors.
   - (d) Using a good vaccination program.
   - (e) Becoming familiar with important diseases and diagnostic lab procedures.
3. Vaccination:
   - (a) Marek's Disease: 1 day (at Hatchery)
   - (b) Bronchitis: 10 days, 4 to 5 weeks, 14 to 16 weeks
   - (c) Newcastle Disease: 2 weeks repeat every 2 to 3 months
   - (d) Fowl Pox: High Risk Area - 1 to 2 weeks, 12 to 16 weeks; Low Risk Area - 8 to 16 weeks
   - (e) Vaccination schedules may vary depending on the disease condition in the area.

Diagnostic Laboratories

There is a charge of diagnostic service. Birds may be sent to these labs by express or by person. Birds should not be shipped to arrive on weekends. Three or four live birds showing typical symptoms of the condition prevalent in the flock should be sent. The shipment should be accompanied by as much information concerning the flock and sick birds as possible such as: age, duration of illness, previous history of illness and vaccination, size of flock, rate of mortality, medication administration, feed intake, production patterns, etc. All birds submitted as well as containers will be disposed of at the lab.

Brooding

Floor houses :

1. Clean and disinfect houses and equipment completely.
2. Put in 4 to 6 inches of clean litre.
3. Have water and feed available when chicks arrive.
4. Inspect and adjust brooders prior to arrival of chicks.
5. Follow manufacturer's directions for brooder capacity.
6. Adjust brooder temperature at edge of hover at 95 degrees F. for the first week and reduce 5 degrees each week.
7. Use brooder guards for first 7-10 days.
8. Isolate brooder house and keep visitors out.

Cage Brooder houses

Many poultry farmers are now brooding pullets in cage brooders. Follow manufacturer's directions on heating systems. Use the same management recommendations for preparing house, feeding, watering and chick care.

Pullet Management :

1. Isolate pullet houses and keep visitors out.
2. Remove unthrifty pullets as they appear.
3. For feeding and lighting details check with an Extension Poultryman.
4. Weigh a sample of birds biweekly to check body weight against strain standard.
5. Debeaking - There are several satisfactory methods and programs of debeaking. The poultryman should select the program which fits his operation. The following methods are used:

- Early Precision Debeaking
  - o Age of bird between 6 and 9 days.
  - o Both beaks inserted into a 10/64 or 11/64 inch hole.
  - o Hot blade (1,700 degrees F) cherry red.
  - o Hot blade is held against beak for 2.5 seconds.
  - o When done properly, one debeaking will last for the expected life of the bird.
- TT Method
  - o Age of bird between 8 and 10 weeks.
  - o Sometimes used in addition to early precision.
- Conventional Method
  - o Age of bird between 12 and 18 weeks.
  - o Independent block cutting of both beaks.
  - o Permanent and gives good protection during laying period.

Layer Management :

1. House only well developed, well fleshed pullets.
2. Use artificial lights to provide 15 hours of total light per day, and start with 1 foot candle intensity at the bird level.
3. Feed a balanced diet - Keep feed intake records.
4. Weigh a sample of birds monthly.
5. Remove obvious culls.
6. Observe birds often to spot problem areas early.

Egg Handling for Quality :

1. Keep cages and egg gathering equipment in good repair to prevent cracks.
2. Handle eggs carefully.
3. Clean dirty eggs using clean, warm (110-120 degrees F) water with a detergent sanitizer.
4. Cool eggs quickly and keep them cool. Hold at a temperature of 50-60 degrees F. with a relative humidity of 70-80%.

Hatching Egg Production and Hatchery Management

Flock Management

Keep birds healthy, use 10 males per 100 broiler breeder females or 8 males per 100 egg breeder females.

Feed

Feed a breeder diet which is specially formulated for high hatchability.

Care of Hatching Eggs :

1. Select eggs for size, shape, shell colour and texture, 23-28 oz. per dozen is best for hatching.
2. Hold eggs at a temperature between 55-65 degrees F., humidity of 70-85% and for not more than 7 to 10 days before setting.

Asia - Biomin again Hosted the Biennial Event, Biomin Asia Nutrition Forum

Biomin, a leading company in the global animal health and nutrition industry, again hosts the biennial event, Biomin Asia Nutrition Forum. The forum catered specifically to the Asian audience, spanning over six cities across Asia from 10 to 21 October, from Cebu to Pattaya, Coimbatore, Chang Sha, Zheng Zhou and Tokyo. The theme of the forum is 'Sustainability: defining the basics, addressing the essentials, introducing NutriEconomics®'.

Professor David Farrell, a key speaker, said: "The great challenge in agricultural science today is to produce sufficient food in a sustainable system that will feed the nine billion people expected to be on this earth by 2050. As the price of food escalates and food riots are happening amongst the poor, we must get smarter to meet the challenge. There are solutions and there will be costs."

With more people to feed and finite resources to depend on, the feed and animal production industry has to grow profitably without compromising the ability to meet future needs.

Targeting at key industry professionals from the poultry, pig, dairy and aquaculture sectors, the forum featured an impressive panel of leading industry experts like David Farrell, Professor from the University of Queensland; John Baize, President of John C. Baize and Associates; Robert Van Barneveld, Professor from the University of New England and Consultant Research Scientist (Nutrition) of Barneveld Nutrition Pty Ltd., and Maximilian Schuh, Consultant and experienced veterinarian in diseases of swine and cattle.

Representing Biomin are Jan Vanbrabant, CEO of Biomin Asia; Guan Shu, Technical Manager of Biomin Asia; Jim He, Technical Nutrition Manager of Biomin Asia; Wolfgang Markert, Director Development Department of Biomin Holding GmbH and Franz

Waxenecker, Director Innovation Management of Biomin Holding GmbH.

Delegates gained insights on various topics that address sustainability from different facets. Presentations explored the solutions to feed the growing population, define the basics of sustainability using the NutriEconomics programme and looked into the implications of commodity trade on animal production.

In addition, there were discussions about feeding the livestock in 50 years, a review of the ban of AGP in Europe and identifying the expectations and trends of antibiotic growth promoter use in Asia. Biomin shared the latest mycotoxins survey results, revealed the future of mycotoxin degradation and unveiled innovations in animal health and nutrition.

Chapter 8

Bacterial Diseases

Escherichia Coli Infections

Omphalitis (Navel Infection)

It is characterised with reddening and tissue oedema in the umbilical region. Escherichia coli infections are widely distributed among poultry of all ages and categories. They are primarily related to poor hygienic conditions, neglected technological requirements or to respiratory and immuno-suppressive diseases. A common sequel of navel infections is local or diffuse peritonitis. When the amount of egg white is bigger (in larger eggs), it impedes the absorption during hatching, resulting in subcutaneous jelly-like oedemas that are an excellent media for the development of E. coli infections.

The delayed absorption of the yolk sac is a prerequisite for E. coli infections and peritonitis. The most commonly identified E. coli serotypes are: 01:K1 (L); 02:K1 (L) and 078:K80 (B).

At a later stage of the infection, the yolk content is a cause of putrefactive necrotic processes in the peritoneal cavity. The abdomen is bloated. The entire abdominal wall is affected by a moist gangrene (maceration).

Salpingitis (Inflammation of the Oviduct)

Salpingites due to E.coli infections could be also observed in growing birds. The oviduct is dilated, with thinned wall and filled with caseous exudate all along its length. Salpingites are among the commonest causes for death in layer hens. E.coli penetrates from the cloaca via an ascendant route. Predisposing factors are the intense egg laying and the associated estrogen activity. In older cases, the caseous masses in the oviduct have a lamellar structure. E.coli organisms are usually found in excreta because of their presence in avian and mammalian intestine, the birds are constantly at risk of infection through contaminated water, dust, faeces and environment.

Retained yolks among the caseous masses in the oviduct. In some cases, when the systemic resistance is lower, places, contaminated with E.coli, such as intestine, genital tract or nasal passages, could be latent sources of infection. Compression and compactedness of caseous necrotic masses after losing a part of their water content in the oviduct of a layer.

Egg yolk peritonitis in a layer hen consequently to E.coli salpingitis. The chickens could be hatched with a latent infection, when E.coli is present in ovaries and the oviduct. In these instances, the infection could turn into an overt infection under the influence of some stress factors or lesions. (Inflamation of the ovary) consequently to a salpingitis due to ascendant E.coli infection. Cystic degeneration of ovarian follicles following an E.coli opphoritis.

Cellulitis

Cellulitis Inflammation of the subcutaneous tissue that affects also the overlying skin. It predominates in broilers and is detected mainly in slaughter-houses. Macroscopically, the lesions are with a yellowish-brown colour. Cellulitis effected areas are mostly in the region of the back and the thighs. In some cases, the lesions are slightly prominating over the adjacent healthy skin. In the subcutaneous tissue, thick fibrinous plaques are often found out. In some cases with adult birds, in the region of the head, subcutaneous masses of thick serofibrinous exudate resulting from a local E. coli infection could be detected.

Enterocolitis

Enterotoxigenic E. coli that produce toxins, cause the secretion and retention of fluids in some intestinal loops and especially in the caeca. Clinically, diarrhoea and dehydration are observed. The intestines are pale and distended, particularly the caeca that are overfilled with fluid containing many gas bubbles.

Neonatal E.coli Septicaemia: Chickens in the first 24 - 48 h after hatching are affected. The death rate during the first ten days is higher and could reach 5-6%. The yolk sac is unabsorbed. The spleen is enlarged. Some days later, the typical serofibrinous polyserositis lesions, affecting the peritoneum, the pericardium, the air sacs and the liver capsule are manifested.

Acute E.coli septicaemia in layer hens. Clinically and morphologically, the acute E.coli septicaemia could resemble fowl cholera or fowl typhoid. It is encountered in both young and sexually mature birds. The stress in the beginning of egg-laying is considered as an important predisposing factor. The parenchymal organs are enlarged and hyperaemic. Sometimes, the liver has a greenish colour and is mottled with multiple small necrotic foci. Also, pericarditis, peritonitis and petechial haemorrhages on serous coats are present.

E.coli septicaemia of a respiratory origin. In such cases, the respiratory mucosa damaged by infectious and non-infections agents (ND viruses including vaccinal strains, IB, TRT, mycoplasmae, high ammonia levels) is the entrance door of the E.coli infection. The lesions are principally observed in the respiratory tract (trachea, lungs and air sacs), but some adjacent serous coats (pericardium, peritoneum) are also affected and thus, the picture of a typical serofibrinous polyserositis is produced.

E.coli septicaemia secondary to enteritis. It is most commonly encountered in turkeys. The intestinal mucosa, damaged by the haemorrhagic enteritis virus, is the entrance door of E. coli infection. The most typical lesions are the marked enlargement, hyperaemia, haemorrhages and necrosis of the liver and the spleen.

Panophthalmitis (Inflammation of All Tissues of the Eyeball)

Generally, it develops secondary to E.coli septicaemia and is usually unilateral. Arthritis, Osteomyelitis and Osteonecrosis (inflammation of joints, bone marrow and bone necrosis, respectively). The lesions are a common sequel to E. coli septicaemia. Clinically, lameness, prolonged lying down, dehydration and retarded growth rate are observed. The coxofemoral joints, the femur and tibiotarsal joints are most commonly affected. The bacteria colonise the physes of growing bones and provoke an inflammatory response that is further causing osteomyelitis. Pathoanatomically, fractures of the femoral head are usually discovered. Some cases of E.coli septicaemia, accumulation of exudate between the superficial and deep pectoral muscles is observed. The lesions that develop in the articular spaces of thoracolumbar vertebrae result in spondylitis (spondylosis) and after that, in progressive paresis and paralysis.

Coligranuloma (Hjarre's Disease)

It is characterised by multiple granulomas in the intestinal tract, the mesentery and the liver, but not in spleen. The lesions are similar to these observed in tuberculosis.

Bursitis Sternalis (Inflammation of the Sternal Bursa)

The bursa is enlarged in a various extent and filled with inflammatory exudate. The diagnosis of coli-infections is based on isolation and typisation of pathogenic E.coli serotypes. Many other bacteria (salmonellae, pasteurellae, staphylococci, etc.), viruses, chlamydiae and mycoplasmae should be excluded as possible aetiological agents. The prevention should aim at minimising the probability of faecal contamination of eggs. This implies the maintenance of clean nests, discarding floor eggs and removal of eggs that are cracked or contaminated with faeces. Breeder eggs should be fumigated or disinfected in the farm prior to their transportation in the storage premise. The treatment is effective if initiated soon after testing the antibacterial sensitivity of isolates.

Pullorum disease is an acute systemic disease in chickens and turkey poults. The infection is transmitted with eggs and is commonly characterised by a white diarrhoea and high death rate, whereas adult birds are asymptomatic earners. The morbidity and the mortality rates increase about the 7th - 10th day after hatching. The affected chickens appear somnolent, depressed and their growth is retarded. The feathers around the vent in many chickens is stained with diarrhoeic faeces or pasted with dry faeces.

The oedema of tibiotarsal joints is a frequent associated sign. Pullorum disease is widely distributed among all age groups of chickens and turkeys. The highest losses are in birds under the age of 4 weeks.

The aetiological agent is S.pullorum, a non-motile Gram-negative microorganism. S. pullorum is very resistant under moderate climatic conditions and could survive for months. It could be killed by fumigation with formaldehyde of breeder eggs in the hatchery. Typical for this form are the greyish-whitish nodes in one or some of the following places: heart (39), lungs, liver, gizzard walls (40) and intestines, the peritoneum.

Sometimes, greyish-whitish milliary necroses are found out in the liver. S.pullorum is transmitted by infected eggs of layer hens that are carriers. Many hatched infected chickens spread the microorganism by a horizontal route to other birds via the gastro-intestinal and the urinary tracts. Adult carrier birds also spread the agent through their excreta. Ureters are often filled with urates. For confirmation of the diagnosis, pullorum should be isolated and typed. Pullorum disease must be differentiated from other salmonelloses, E.coli infections, Aspergillus that produces similar pulmonary lesions. Staphylococcus aureus, causing arthrites, etc. Sometimes, the pulmonary nodes resemble the tumours in Marek's disease.

Acute Fowl Typhoid

The outbreaks usually begin with a sharp decline in forage consumption and egg production. The fertilization and hatchability rates are considerably reduced. Diarrhoea appears. The death rate in acute fowl typhoid is high and varies between 10% and 90%. About 1/3 of chickens hatched from eggs from typhoid-infected flocks die. A characteristic lesion for acute fowl typhoid in adult birds is the enlarged and bronze greenish tint of liver.

In some instances, the enlarged liver is mottled with multiple milliary necroses. The outbreaks are observed primarily in hens and turkeys, but the disease is sometimes encountered in other domestic or wild fowl. In other cases, the size of liver necroses varies from milliary to spots with a diameter of 1-2 cm. Unlike pullorum disease, fowl typhoid is lasting for months. Acute fowl typhoid. The spleen is 2-3 times bigger, sometimes with greyish-whitish nodules prominating on the surface, representing hyperplasic follicles. Often, enteritis, especially of the anterior part of small intestine, sometimes with ulcerations, is present. The aetiological agent is Salmonella gallinarum. This organism usually shares common antigens with 5. pullorum and the two microorganisms often give a cross-agglutination reaction.

More rarely, myocardial necroses due to Salmonella toxins are detected. The transmission of the infection by contaminated eggs is especially important. Moreover, the transmission of 5. gallinarum occurs mainly among growing or productive flocks and the death rate among adult birds is higher. The lungs acquire a characteristic brown colour. Here, necroses and, following their organisation, "sarcoma-like nodules"

Chronic Fowl Typhoid

The lesions are primarily in the gonads. The ovaries are affected by inflammatory and degenerative changes. Frequently, affected follicles are deformed and appear like thick pendulating masses. Fowl typhoid should be differentiated from other salmonelloses, E. coli infections, Pasteurella spp. infections, etc. If breeder flocks are proved to be carriers of the infection, their eggs should not be used for breeding.

Sometimes, the going out of yolk from degenerated follicles results in fibrinous adhesive peritonitis. Taking into consideration that chemotherapy does not eliminate the carriership, the treatment of poultry infected with fowl typhoid or pullorum disease is not justified and is never recommended.

Riemerella Anatipestifer Infections

Riemerella anatipestifer (RA) infection is a contagious disease in domestic ducks, turkeys and other fowl. It is encountered as acute or chronic septicaemia and is characterised with serous fibrinous polyserosites. The respiratory tract could be also affected. The ducklings at the age of 18 weeks are especially susceptible. Clinically, sneesing, cough, trembling of the head and neck, ataxia and greenish diarrhoea could be present.

The most characteristic gross lesion is the deposit of fibrinous exudate on the pericardium, the liver capsule or air sacs. The chronic lesions affect the skin and the joints. Although a tentative diagnosis could be made on the basis of observed clinical symptoms and lesions, it is confirmed upon the isolation and identification of RA. The RA infection should be distinguished from septicaemiae due to P. multocida, E.coli, Salmonella, etc. The treatment with antibiotics (Flumequine) and sulphonamides IJrimetoprim, Sulphadiazine) has a varying success.

Staphylococcus aureus infections are commonly seen in poultry. Usually, bones, tendon sheaths and joints, particularly coxofemoral and tibiotarsal joints are affected. The clinical signs include unilateral or bilateral lameness, reluctance to move and lying down. When the tibiotarsal joints are affected, swellings, fever and sometimes necroses of overlying tissues and purulent exudation are observed. As a sequel to septicaemia, osteomyelitis could occur. The lesions are usually detected in the region of the proximal femur, where inflammatory necrotic foci in the bone marrow and partial or complete fracture of the femoral head are observed.

The outcomes of skin injuries are the appearance of cellulites, characterised by extensive purulent inflammation of the subcutaneous tissue. Most staphylococcal isolates are identified as Staphylococcus aureus, Gram-positive cocci, arranged in clusters. Plantar abscesses are a kind of local purulent inflammation. They are located on the plantar surface or adjacently to it and appear following puncture wounds. Staphylococci are moderately resistant to common disinfectants. Chlorine-containing disinfectants are efficient in the absence of organic matter.

Other local forms of staphylococcosis are sternal bursites. The sternal bursa is enlarged at a various extent because of gathering of purulent substance. Sometimes, the covering skin is necrotised. The microorganism is widely distributed in the environment and mainly on the skin. Most of Staphylococcus oureus-induced lesions are associated with skin injuries, debeaking, finger cuts. All categories of birds are affected.

Toxicogenic strains are able to induce food intoxications. In a septicaemic staphylococcal infection, hyperaemia, enlargement and various-degree coagulation necroses in the liver or the spleen are observed. As staphylococci are ubiquitous, their presence could not be prevented. The measures should be directed towards minimising the possibilities for traumas of skin, respiratory and intestinal mucosa.

A typical secondary infection is staphylococcal gangrenous dermatitis that is usually resulting from immunosuppressive infections (IBD, CIA). Affected skin areas are dark red to blue-greenish, moist and sharply defined from the adjacent healthy skin. Histologically, staphylococcal lesions consist of necroses, bacterial colonisation and heterophilic infiltration. The isolation and identification of Staphylococcus spp. confirms the diagnosis. From a differential diagnostical point of view, infections with E. coli, S. gallinarum, P. multocida, M. synoviae and reoviruses should be considered.

Streptococcosis in poultry is encountered as acute septicaemia or chronic infections with a death rate between 0.5 and 50%. The agent that is most commonly associated with several pathological conditions in poultry is 5. zooepidemicus. The lesions in chronic streptococcal infections include arthrites, tenosynovites, myocardites and valve endocardites. Endocardites affect predominantly mitral valves and less frequently, aortal and tricuspid valves.

Other lesions related to endocardites, are heart, liver and spleen infarctions. Liver infarctions are usually peripherally located on margins, have a pale creamy colour and are sharply demarcated. The penetration of the infection occurs mainly via the oral or aerogenic route, but could also enter through the injured skin, especially in battery cage layer hens. The differential diagnosis includes other bacterial septicaemias as staphylococcosis, fowl cholera, E. coli infections, etc.

Mycoplasma

MG is characterised by respiratory symptoms and a prolonged course of the disease. Particularly susceptible are hens and turkeys at all ages. The aetiological agent is M. gallisepticum. In many cases however, the pathogenicity of the microorganism is enhanced because of its association with any or some of the following agents: E. coli, P. multocida, H. paragallinarum and IB or ND viruses. The most characteristic signs in adult flocks are tracheal rales, nasal discharge, coughing, decreased egg production. Most outbreaks are in broiler chickens older than 4 weeks. The course of the disease is more severe during the winter and in cases of associated infections. Often, conjunctivrtes, facial skin oedema and profuse tear secretion could be observed. In turkeys, unilateral or bilateral swelling of periorbital sinuses, nasal discharge and conjunctivitis are observed. The inflammatory exudate is commonly fibrinous and is detected as diffuse accumulation after removal of overlying skin.

Most common gross finding is aerosaccculftes, the air sacs being filled with fibrinous caseous exudate. The majority of routine chemical disinfectants are effective against M.gallisepticum that rarely survives longer than a few days away from the host. The birds could carry the microorganism and be asymptomatic until the disease is triggered by stress factors such as change of the premise, the diet or weather, vaccinations against or infections with IB or ND, increased levels of dust or ammonia.

In older cases, the content of air sacs is dense and compact. A vertical transmission is done through the eggs of some unapparent carriers. The infected progeny transmits the agent horizontally via airborne route, by coughing or contaminated forage, water and environment. Serofibrinous pneumonias, usually bilateral, are a frequent finding.

Often, the inflammation involves the adjacent serous coats and thus, fibrinous polyserosites occur. Sinusites are relatively rarely observed in hens. The positive agglutination tests of sera in several birds from the flock confirm the diagnosis. MG should be distinguished from other respiratory diseases in poultry. Pulmonary and air sacs lesions could be mistaken with similar findings in E. coli septicaemia or aspergillus's. In turkeys, P. multocida pneumonia should also be considered.

Mycoplasma synoviae (MS) infections could progress as either acute or a chronic systemic disease with symptoms of arthritis, synovitis and bursitis especially in hens and turkeys. The earliest signs are lameness, lying down and retarded growth. Often, oedemas of tibiotarsal joints and the drumstick are observed. The morbidity and death rates are moderate, under 10%. Young chickens at the age of 4-12 weeks and turkey poults at the age of 10-12 weeks are susceptible. Synovites are encountered all year round, but are prevalent during cold humid seasons or when the litter is wet.

Affected birds get progressively exhausted. When the joints and tendon sheaths are open, a serofibrinous exudate is most commonly observed. The aetiological agent is M. synoviae. The microorganism shows a certain tropism to synovial structures as joints and tendon sheaths. An important route of dissemination of the agent is the transovarial transmission. The distribution by a horizontal route via the respiratory tract is also possible. The commonly used means of diagnostics is ELISA. MS infections should be differentiated from staphylococcal infections, reoviral arthritis and RGT.

Necrotic Enteritis

Necrotic enteritis (NE) is an acute Clostridium infection characterised by severe necroses of intestinal mucosa. The disease begins suddenly, with a sharp increase in death rate. A strong dehydration is observed. The skin is sticked on or adhered to body musculature and is hardly removed. Chickens at the age of 25 weeks are usually affected, NE is also encountered in hens particularly near the period of the beginning of egg laying or peak egg laying, most commonly associated with coccidiosis. In acute cases, marked congestion of liver, responsible for its dark red to black appearance, is present.

The aetiological agent is Clostridium perfringens, mainly from type A and more rarely from type C. The produced a and p" toxins, from C. perfringens type A and type C respectively, are responsible for the necrosis of intestinal mucosa. The small intestine is often distended with gases and the necrotic mucosa is visible through the wall. CI. perfringens is ubiquitous and normally reside into the intestinal tract. The alterations are particularly in the jejunum and the ileum because of their higher pH and the lower oxygen content in these areas. Sometimes, haemorrhages are seen through the intestinal wall. The intestinal lumen is filled with brownish watery content, mixed with gas bubbles. Necrotic mucosa acquires a greyish-creamy or greenish appearance. Sometimes the mucosa has a flannelette blanket-like appearance.

In some cases, the mucosa has a linear pattern similar to the bark of a tree. The predisposing factors are injuries of intestinal mucosa by various Eimeria species, migration of ascarids, immunodeficiency states due to CIA, IBD, MD, high content of wheat or fish meal in the diet.

In cases when NE is associated with small intestinal coccidioses, multiple petechial haemorrhages could be perceived through the wall in different areas along the small intestine. Throughout the simultaneous occurrence of NE and coccidioses, the content of the lumen is bloody, mixed with necrotic detritus and gas bubbles. The diagnosis is based on the distinctive gross lesions. When necessary, a histological investigation is performed or attempts for isolation of the causative agent. NE should be distinguished from ulcerative enteritis and some small intestinal coccidioses. The control should be aimed at predisposing factors. An appropriate medication of feeds is recommended. A good effect is obtained with oxytetra-cycline dihydrate (OTC 50% premix). NE could be effectively treated with doxy-cycline hydrochloride, amoxicillin, etc.

Cholangiohepatitis (CAH) in broiler chickens is characterised by inflammatory proliferative and dystrophic necrobiotic alterations in bile ducts and the liver parenchyma. Usually, no clinical signs are observed. The increased daily mortality is insignificant, although in some chickens, a retarded growth and dehydration could be present. Pathoanatomically, the liver is enlarged and with paler yellow colour. In some cases, its surface has a characteristic acinous appearance and in others is mottled with multiple small greyish-white or greenish foci.

The walls of the gall bladder are thickened, sometimes up to 5-6 cm, and opaque. The state is detected in the last phase of the fattening period or in the slaughterhouse. It is possible to observe CHA as an independent disease or associated with necrotic enteritis. A transverse cross section through the thickened gall bladder wall. Clostridium perfringens is the aetiological agent. CAH is experimentally reproduced in broilers by ligation of bile ducts and inoculation with CI. perfringens. The gall bladder is filled with a thick bile secretion or a dense matter with a creamy colour.

In some chickens, the subcutaneous fat and the body fat have an icteric tint. Histologically, the liver lesions are detected at a various stage of development. In the majority of cases, proliferative changes in bile ducts are observed. The overgrown bile ducts form granulomatous structures, surrounded by fine reticular fibres. Centrally, in some granulomas, either initial or advanced degree of necrosis and weak to moderate granulocytic infiltration are observed.

In many bile ducts, a biliary stasis is present and within some, among the stagnated secretion, a huge amount of microorganisms are detected. Pericanalicularly, coagulation necroses are frequently noticed. Among some of these foci, hyalinisation of the necrotic masses and single microorganisms are present. Around the necroses, a belt of macrophages, lymphocytes and granulocytes is formed. Many Gram-positive bacteria are detected among granulomas, in bile ducts' lumen and into the gall bladder, often accompanied by inflammatory lesions.

The mucosa of bile ducts and the gall bladder is frequently necrotised and the wall is thickened because of connective tissue growth. The diagnosis is based on the characteristic gross and microscopic lesions. The prevention is similar to that in NE. Ulcerative enteritis (UE) is characterised by inflammatory ulcerative and necrotic changes in intestinal mucosa and dystrophic necrobiotic lesions of the liver and the spleen. The clinical signs include a general malaise, ruffled feathers, diarrhoea and anaemia. In many instances, the disease begins with a sudden death. Pathoanatomically, deep button¬like ulcers are observed, mainly in caeca and less frequently, in some parts of the small intestine, usually visible through the wall.

UE is a problem in all world regions with extensive poultry breeding. Young birds are infected more frequently although the disease is also common among adult quails. The early lesions appear like yellowish foci with haemorrhagic boundaries that could be seen from both the serous and the mucosal surfaces.

The aetiological agent is Clostridium colinum, a spore-forming organism, highly resistant to chemical agents and physical alterations. The intestinal content is often mixed with blood.

In older and larger ulcers, the haemorrhagic zones tend to disappear. The ulcers could have an irregular round or elongated shape and are covered by large necrotic diphtheritic membranes.

Frequently, adhesive peritonitis due to inflammatory involvement of adjacent serous coats is observed. Numerous domestic and wild birds (chickens, quails, turkeys, rock partridges, geese, partridges, etc.) are susceptible. The chickens and the quails are the most vulnerable between 4 and 12 weeks of age where as turkeys between 3 and 8 weeks of age. Liver, a variety of dystrophic changes and necroses with different size and shape are detected. Necrotic foci in some cases are milliary. Sometimes, liver necroses reach 1 - 2 cm. in diameter and are surrounded by a haemorrhagic zone.

In some instances, necrotic foci involve large parts of the liver and are infiltrated by haemorrhages. The aetiological agent is distributed with the excreta of acutely ill and recovered birds and persists in soil for many months. The incubation period is 1 - 3 days. The death rate in chickens varies from 2% to 10% and in quails reaches 100%. The outbreaks of UE in chickens are often associated with or come after coccidioses, CIA, IBD or stress conditions. Most commonly, necroses are distinguished on the background of a marked parenchymatous dystrophy, affecting partially or totally (uni- or bilaterally) the liver. The spleen could be enlarged, hyperaemic, haemorrhagic and sometimes, with necroses. The diagnosis is based on the typical gross lesions. When needed, imprint preparations are made, a histological study is performed or attempts for isolation and identification of the aetiological agent are made. UE should be differentiated from NE, coccidiosis and histomonosis (typhlo-hepatitis).

Some cases, haemorrhages with various intensities are detected in the mucous coat of the gizzard. Prevention -separate housing of the different age groups of birds, avoiding the contact with other avian species. The pre-medication of forages with some antibiotics and their rotation would prevent the reproduction of CI. colinum. A good effect is achieved with oxytetra-cycline dihydrate (OTC 50% premix). UE could be effectively treated with doxycycline hydrochloride, amoxycillin, etc.

Gangrenous Dermatitis

The gangrenous dermatitis (GD) is a disease affecting fattening and growing birds, characterised with necrotisation of different skin areas and a severe cellulitis of the subcutaneous tissue. The sudden and quick increase in death rates is often the first signal for the incidence of GD. The affected birds die after less than 24 h. The death rate is from 1% to 60%. The lesions are dark red to blue green macerated skin areas, usually featherless, beginning generally from wings and the adjacent areas. Most outbreaks are encountered in broilers at the age of 4-8 weeks. The disease is also observed in stock layer hens at the age of 6-20 weeks and broiler parents at the age of 20 weeks. The outbreaks are frequently observed in extremely wet and warm premises.

The gangrenous dermatitis affects birds while still alive. In more s evere cases, the gangrene could begin and involve the skin of the head, neck and the breast. Affected skin is macerated or totally necrotic, resulting in exposure of underlying tissues in a number of cases.

The GD agents are CI. septicum, CI. perfringens type A and Staphylococcus aureus, independently or in combination. The associated infection is more severe. The skin lesions are often crepitating and are detected in the regions of breast, abdomen, back or wings in both alive and dead birds.

Under the affected skin, extensive haemorrhagic oedemas with or without gas (emphysema) are discovered. The Clostridia are encountered in soil, faeces and the intestinal content. Staphylococci usually inhabit the skin and mucous coats of birds Underlying muscles are haemorrhagically infiltrated and also could contain gas among the muscle fibres. The increased susceptibility of birds to GD is related to immunodeficiency states secondary to CIA, IBD, IBH, etc. Other predisposing factors are aflatoxicoses, the unbalanced or deficient diet or poor hygiene. The skin wounds are the entrance door of the infection.

In most cases, no changes in viscera are observed. Rarely, gas bubbles could be found out in the liver. The diagnosis is confirmed by imprint preparations or histological cross sections. The bacterial agent could be isolated from cellulitic zones.

Single or multiple necroses are rarely seen in the liver. Prevention - the factors leading to skin injuries (cannibalism, mechanical traumas, etc.) should be minimised. Successful approaches are the immunoprophylaxis programmes in breeder flocks aiming at prevention and reduction of possible immuno¬suppressive states. Microscopic lesions are characterised by oedema, emphysema, hyperaemia, haemorrhages and necroses in the subcutaneous tissues. GD could be effectively treated by tetracyclines: doxyxycline hydrochloride, erythromycin thiocyanate, chlortetracycline and copper sulphate in the water.

Botulism

Botulism is an intoxication caused by the toxins of Clostridium botulinum. With the exceptions of vultures, most birds are susceptible. The clinical signs appear within a few hours to several days. Flaccid paresis of legs, wings, necks and eyelids is observed. The paresis is rapidly progressing to paralysis and the birds fall into a deep coma with neck and head typically extended forward. There are no typical gross lesions. The treatment with selenium, vitamins A, D and E as well as with some antibiotics as chlortetracycline, bacitracin, etc. could reduce mortality rate.

Spirochaetosis is a septicaemic disease characterised by depression, progressive paresis, paralysis and inflammatory necrobiotic changes in parenchymal organs and the gastro-intestinal tract. The birds are depressed, cyanotic; greenish diarrhoea with considerable amounts of urates is observed. The transmission of spirochaetosis is related to the distribution of the fowl tick Argas persicus that is the reservoir and main vector. Hens, turkeys, geese, ducks, pheasants, etc. from all ages are susceptible. At a later stage, paresis and paralysis are developing, the birds become somnolent and comatose.

After removal of plumage, the ticks, adhered to the skin, are discovered. The aetiological agent is Borrelia anserina, a highly motile spirochaete with 5-8 coils. The organism is not resistant outside the host and could exist only in some vector. A typical finding for spiro¬chaetosis is the enlarged spleen with marble-like appearance. A. persicus remains infective for over 430 days and transmit spiro-chaetes to its progeny. Spirochaetes could be identified in Giemsa-stained blood smears. In the late stage of the disease, spirochaetes could not be detected.

The liver is often enlarged, mottled with necroses of a various size and marginal infarctions. Usually, bile-coloured mucoid enteritis is observed. The arsenics as well as some antibiotics including tylosin, tetracycline and penicillin are effective for treatment of infected birds. The application of vaccines is successful, but the acquired immunity is short and revaccinations are needed.

Avian Tuberculosis

Avian tuberculosis is a chronic infectious disease characterized by the formation of granulomatous lesions in viscera, a progressive weight loss and death. It is usually encountered sporadically in birds reared in small yards, zoos and is a problem among caged exotic birds.

The aetiological agent is Mycobacterium avium, a very resilient and acid-resistant microorganism. It is resistant to temperature changes, drying, pH changes, to many disinfectants and survives in the soil for years. Along the small intestine, single or multiple sub-serously prominating tubercles are detected. In advanced cases, tubercles could be observed along the bone marrow of the femur or the tibia. The main mass of the tubercle is composed of epitheloid cells, and peripherally, there are foreign body-type multinucleated giant cells. In older granulomas, central coagulation or caseous necroses are present.

Characteristic features of avian tuberculosis are conglomerate tubercles (gross and histological appearance). The diagnosis is based upon the complex evaluation of history, persisting lethality in adult flocks and the pathoanatomical findings. Avian tuberculosis should be differentiated from neoplastic diseases, coligranu-loma (Hjarre's disease), pullorum disease, etc. The treatment is not advised, as the disease is contagious for men.

Viral Inclusion Body Hepatitis

The viral inclusion body hepatitis (IBH) is an adenovirus infection characterised by haemorrhages and dystrophic necrobiotic changes in the liver and kidneys, accompanied by intranuclear inclusion bodies. A characteristic macroscopic lesion is the enlarged, dystrophic liver with yellowish colour and crumbly texture.

IBH outbreaks are encountered primarily in meat type chickens, most commonly at the age of 3-8 weeks. IBH often occurs as a secondary infection to immunodeficiency resulting from other diseases (IBD, CIA). On the background of dystrophic liver changes, haemorrhages of various intensity and size are outlined, thus creating a variety of liver lesions. Throughout IBH outbreaks, several serotypes from the 12 known avian adenoviruses (AAVs) of group I are isolated. The sick chickens cany the virus in their excreta, kidneys, tracheal and nasal mucosa. The virus is resistant to many environmental factors and could be easily transmitted by a mechanical route. The transmission of adenoviruses is realised vertically by breeder eggs and horizontally, via excreta (mainly faeces). In a number of cases, the dominating lesion is the massive mottled or striated haemorrhages of the liver.

IBH is characterised by a sudden onset and a sharply increased death rate that reaches peak values by the 3rd - 4th day and returns back within the normal range by the 6th - 7th day. The total death rate is usually under 10% but sometimes could attain 30%. More rarely, macro-scopically visible necrotic foci could be detected in the liver.

The kidneys are enlarged, pale and mottled with multiple haemorrhages. Many instances of IBH, the amount of pericardial fluid is increased (hydropericardium).

Clinical signs could be observed only several hours prior to death occurrence. They consist in pale comb and wattles, depression and apathy. Sometimes, the skin is icteric. Often, ecchymoses and striated haemorrhages in skeletal muscles are observed.

Microscopically, extensive dystrophic changes and necroses of liver parenchyma are detected. In the nuclei of hepatocytes, basophilic or eosinophilic inclusion bodies are detected. The basophilic inclusion bodies are usually dense and occupy the entire nuclear inner space (arrow b), whereas the eosinophilic ones are round or irregularly shaped and surrounded by a light halo (arrow e). The diagnosis is based upon the typical gross lesions and the history records. A principal approach in IBH diagnostics is the histological investigation that helps to detect the intranuclear inclusion bodies. IBH should be distinguished mainly from IBD and chicken infectious anaemia (CIA). With regard to IBH prevention and control, the eggs of broiler parent flocks, where the disease is consecutively appearing in the progeny, should not be used for hatching. The access of wild birds should be prevented as they are potential carriers and distributors of the virus. The most important steps in IBH prevention are the control of IBD and CIA. There are neither vaccines, nor an effective treatment.

Haemorrhagic Enteritis of Turkeys

The haemorrhagic enteritis (HE) is an acute viral disease in young turkeys, characterised by a sudden onset, bloody faeces and various, but often high death rate. Blood discharge from the vent, fresh blood in faeces or melena could be observed. HE is observed in turkeys at the age of 6-11 weeks but most commonly at the age of 7 - 9 weeks. As an exception, HE could be observed in turkeys under 4 weeks of age, presumably due to maternal antibodies. The small intestine, especially the duodenum, has a dark red colour and ramiform blood vessels prominating under the serous coat, and sometimes, haemorrhages are seen through the intestinal wall. HE is characterised by rapidly progressing clinical signs and disease course of about 10 days. The peak death rate is reached by the middle of the period. The entire death rate varies between 2-3% and 5-10% but could reach 60% as well. The turkeys are depressed, a sharp reduction in forage and water consumption is seen. The extent of lesions of the interstitial mucosa varies from hyperaemic to haemorrhagic, severe haemorrhagic or fibrinous necrotic inflammation.

The spleen of infected birds is typically enlarged, haemorrhagic, crumbly, motted or marble-like. The aetiological agent of HE is an adenovirus of group (type) II (HEV). The viruses or marble spleen disease in pheasants (MSDV) and avian adenvirus splenimegaly (AAS) that are serglogically similar, beling to the same group.

Later, the spleen reduces its size 2-3 times and acquires a specific silvery-grey colour. The serological studies show that HEV is widely distributed among adult turkeys. There is no evidence for vertical transmission. HEV is spread horizontally by ingestion of faeces of infected turkeys. Contaminated litter, equipment, shoes, other birds, rodents, etc. are also important in the transmission of the infection.

The liver is enlarged, crumbly and mottled with multiple haemorrhages, varying from petechiae to ecchymoses. Dead birds are in a good condition, with pale skin. Once having appeared in the farm, the infection is often recurring in other flocks. The HEV infection in turkeys results in a transient immunosuppression and consequently, a secondary E. coli septicaemia Less frequently, extensive necrotic foci, sometimes haemorrhagically infarcted, could be detected in the liver. The kidneys are enlarged and haemorrhagic. In many instances, the small intestinal and gastric serous coats are encompassed by petechial or striated haemorrhages. Hyperplasia of the white spleen pulp, necroses, subcapsular and periarteriolar haemorrhages are observed.

A characteristic diagnostic sign is the discovery of large acidophilic, rarely basophilic intranuclear inclusion bodies in the reticuloendothelial cells. The displaced and condensed nuclear chromatin pattern around the inclusion bodies often looks like a crescent. Similar inclusion bodies could be sometimes found out in lamina propria of the intestinal mucosa.

The staining of peripheral blood smears reveals a picture, characteristic for septicaemia. The typical gross lesions and the history of the disease allow making a tentative diagnosis for HE. The histological detection of intranuclear inclusion bodies in the reticuloendothelial cells of the spleen or intestines confirms the diagnosis. For identification of the HEV, the agar gel precipitation method could be used.

The viral antigen could be detected in fresh or frozen spleen tissue, diluted 1:1 with saline or in sera obtained from diseased turkeys. HE should be distinguished from other cases of severe enteritis in turkeys, acute E. coli septicaemia and some other septicaemic states (streptococcosis, fowl cholera, etc.). Within weeks after the beginning of the disease, an appropriate antibiotic therapy should be initiated for prevention of secondary E. coli septicaemia. Practically, there is no effective therapy for HE in turkeys.

Viral Diseases

Egg Drop Syndrome

The egg drop syndrome - 1976 (EDS 76) is an infectious disease in layer hens manifested by a quick drop in egg production, failure to reach peak production, irregularly shaped eggs, soft-shelled or shell-less eggs and depigmentation. The aetiological agent is an adenovirus of group III. The horizontal transmission occurs slowly in battery systems and rapidly in floor housing systems. The first sign is the loss of egg pigmentation, rapidly followed by the appearance of soft-shelled, shell-less of deformed-shell eggs. If defective eggs are discarded, the remaining ones have no problem with fertilization and hatching.

The drop could be sudden or prolonged. Usually, it lasts for 4-10 weeks and the egg production is reduced by about 40%. Apart the inactive ovaries and oviduct atrophy, other lesions are not discovered. The replication of the virus in epithelial cells of oviduct glands results in severe inflammatory and dystrophic changes in the mucous coat. The appearance of eggs with impaired quality and the dropped egg production are suggestive for EDS 76.

The diagnosis is supported by some serological studies and is confirmed after isolation and identification of the virus. In many instances, no antibodies are detected in infected flocks until egg production approaches levels between 50% and peak production.

Adenovirus Group I - Associated Infections

Necrotic pancreatitis in broilers. It is characterised by focal necroses in the pancreas. Gizzard erosions in broilers. They are manifested by erosions affecting the cuticle and the underlying tissues of the gizzard. Avian adenovirus splenomegaly. Avian adenovirus splenomegaly is characterised by enlargement of the spleen, pulmonary oedema and congestion. It is observed in broiler breeders at the age of 20-45 weeks. Its course is peracute or acute. The mortality could reach 8 - 9%. The most typical lesions are splenomegaly, mottled or marble-like appearance of the spleen, oedema or hyperaemia of lungs.

Infectious Bursal Disease (Gumboro)

Infectious bursal disease (IBD, Gumboro) is an acute, highly contagious viral infection in chickens manifested by inflammation and subsequent atrophy of the bursa of Fabricius, various degrees of nephroso-nephritis and immunosuppression. Clinically the disease is seen only in chickens older than 3 weeks. The feathers around the vent are usually stained with faeces containing plenty of urates.

The period of most apparent clinical symptoms and high death rate is at the age of 3 - 6 weeks. IBD could however be observed as long as chickens have a functioning bursa (up to the age of 16 weeks). In chickens younger than 3 weeks, IBD could be subclinical, but injured bursa leads to immunosuppression. Also, diarrhoea, anorexia, depression, ruffled feathers, especially in the region of the head and the neck are present.

A natural IBD infection is mostly observed in chickens. In turkeys and ducks it could occur subclinical^, without immunosuppression. Most isolates of the IBD virus in turkeys are serologically different from those in chickens. In premises, once contaminated with the IBD virus, the disease tends to recur, usually as subclinical infection. The dead bodies are dehydrated, often with haemorrhages in the pectoral, thigh and abdominal muscles.

The IBD virus belongs to the Birnaviridae family of RNA viruses. Two serotypes are known to exist, but only serotype 1 is pathogenic. The virus is highly resistant to most disinfectants and environmental onditions. In contaminated premises, it could persist for months and in water, forage and faeces for weeks. The incubation period is short and the first symptoms appear 2-3 days after infection. The lesions in the bursa of Fabricius are progressive. In the beginning, the bursa is enlarged, oedematous and covered with a gelatinous transudate.

IBD virus has a lymphocidic effect and the most severe injuries are in the lymph follicles of the bursa of Fabricius. Most commonly, IBD begins as a serous bursitis.

IBD lesions undergo various stages of serous haemorrhagic to severe haemorrhagic inflammation. The morbidity rate is very high and could reach 100%, whereas the mortality rate: 20 - 30%. The course of the disease is 5-7 days and the peak mortality occurs in the middle of this period. In some cases, the bursa is filled with coagulated fibrinous exudate that usually forms casts with the shape of mucosal folds. In birds surviving the acute stage of the disease, the bursa is progressively atrophying. Microscopically, an atrophy of follicles into the bursa is observed secondary to inflammatory and dystrophic necrobiotic alterations.

The kidneys are affected by a severe urate diathesis. In an acute outbreak and manifestation of the typical clinical signs, the diagnostics is not difficult. The diagnosis could be confirmed by detection of typical gross lesions throughout a patho-anatomical study. IBD should be differentiated from IBH (inclusion body hepatitis). The application of live vaccines in chickens is a key point in the prevention of IBD and should be related to the levels of maternal antibodies.

Infectious Bronchitis (IB)

In chickens up to the age of 4 weeks, IB is manifested by severe respiratory signs (sneesing, coughing, and rales). Rhinites and conjunctivites, depression and crowding around heat sources are observed. The death rate could reach 100%. The mortality in young chickens is usually insignificant provided that a secondary infection with a different agent is not occurring. In such cases, there is a moderate to severe inflammatory cell infiltration of upper respiratory tract mucosa, resulting In thickened and more compact mucosa.

In layer hens infected with the IB virus, oophorites and dystrophic necrobiotic lesions affecting primarily the middle and the last thirds of oviduct's mucous coat are observed. The consequences are drop in egg production, appearance and increase in the number of deformed and pigmentless eggs or eggs with soft shells and watery egg white.

The nephrotropic strains of the IB virus cause severe inflammatory and dystrophic necrobiotic damages of kidneys: urolithiasis, interstitial nephritis, haemorrhages (that considerably increase the death rate. Under natural conditions, only hens are infected. Non-immune birds of all ages are susceptible.

The disease is even seen in vaccinated flocks. The serological methods (VN, ELISA, etc.) are widely used in the diagnostics. At present, PCR is used for rapid identification of IB virus serotypes. IB should be distinguished from other acute respiratory disease as ND, laryngotracheitis and infectious coryza. The vaccination with live or killed vaccines is effective only if they contain the respective serotype of the virus for the given region.

Laryngotracheitis

Laryngotracheitis (LT) is a viral infection in hens, pheasants and peacocks characterised by catarrhal haemorrhagic to fibrinous inflammation of the respiratory tract. It is manifested in laryngotracheal and conjunctival form. In the laryngotracheal form, suffocation, rales and cough are observed. The head and the neck are strongly extended forward and upward during inspiration. Mucous coats of the larynx and the trachea are catarrhally haemorrhagically to fibrinously inflamed. Most outbreaks are encountered between the age of 4 and 14 weeks although the disease affects fowl of any age. LT is caused by a herpesvirus that is relatively resistant. Haemorrhagic laryngotracheitis. The morbidity rate of LT reaches 50-70% and the death rate: 10-20%. Often, it goes on as a complicated infection after the involvement of E. coli, St. aureus, M. gallisepticum, etc. In some cases, casts of haemorrhagic or fibrinous exudate are formed that could almost completely obturate the larynx and the trachea. Source of the infection are sick and convalescent birds, the latter being prolonged carriers of the virus (up to 1 - 2 years). With this regard, a certain stationarity is observed.

In the conjunctival form of LT, wet eyes, tear secretion and oedema of infraorbital sinuses are observed, especially in a complicated infection. The typical clinical and morphological signs are sufficient to assume the presence of LT. The diagnosis is confirmed with the detection of intranuclear inclusion bodies in the trachea throughout the histological study in the early stages of the disease, serological studies (VN, ELISA), etc. LT should be differentiated from IB, SHS, M. synoviae infections, etc. Premises, contamined with the LT virus, should be freed, cleansed, disinfected and occupied again after 5 - 6 weeks. The vaccination of unaffected birds and these from other premises of the infected farm could protect and stop subsequent outbreaks.

Swollen Head Syndrome

Swollen head syndrome (SHS) is a complicated infection in broilers and broiler breeders, where the primary aetiological agent is an avian pneumovirus (APV) and the secondary -usually E. coli; it is characterised by respiratory and nervous signs. In broilers chickens, SHS is generally seen after the 4th week of life. The first clinical signs are sneesing, coughing, rales and conjunctivites. A profuse tear secretion, reddened conjunctivas and a characteristic oblong almondlike shape of eyes are observed. The inflammatory exudate is initially transparent, but afterwards becomes opaque. Subcutaneous oedemas in the head region, involving unilaterally or bilaterally the periorbital sinuses and the mandibular space, are emerging.

A seasonal pattern in the prevalence with peaks during spring and summer is observed. After removal of the covering skin, deposits of serofibrinous exidate are observed. APV is a virus from the Paramyxoviridae family. The pneumovirus is present in respiratory secretions and discharges. It survives for a long time in cold and wet environment. E. coli follows the APV infection of the upper respiratory tract. The triggered inflammatory response results in accumulation of exudate in the subcutaneous tissue. In many instances, a croupous pneumonia develops at a later stage consequently to contamination with other pathogens.

SHS in broiler breeders is usually encountered around or after the peak egg laying period only in female birds. Unilateral or bilateral swellings of the head, affecting the periorbital sinuses, the mandibular space and the wattles are seen. The conjunctiva and the mucous membranes of sinuses are considered to be the entrance door of the infection. A unique cytotoxin has been identified in many E. coli isolates in SHS that could be involved in the pathogenesis of the disease. The inflamed by the virus infection conjunctiva-associated lymphoid tissue is the site where the bacterium enters the subcutaneous tissue.

Frequently, nervous signs are observed in broiler breeders (opisthotonus, torticolis) due to inflammatory processes in pneumatic skull bones and the middle ear. In laying hens, the ovaries are affected in many instances (serofibrinous oophoritis), resulting in reduced egg production. The diagnosis is based upon the distinctive clinico-morphological signs. SHS should be differentiated from Mycoplasma and Pasteurella infections and the skin form of aspergillosis. Prevention - improvement of the microclimate of premises, use of live and killed vaccines.

Infectious Encephalomyelitis

The infectious encephalomyelitis (IEM) is characterised by signs of ataxia, progressing to paralysis, prostration and marked tremor of the head and the neck, and because of that, is also called epidemic tremor. The chickens with prostration are usually in lateral recumbency. The tremor could be unapparent, but is often perceptible when the chicken is held gently with the hand and carefully looked at. the expression of the eye is dull. IEM outbreaks are generally observed in chickens at the age between 8 - 9 and 20 days. the morbidity rate could reach 40 - 60%. the average death rate is about 25%, but could be more than 50%. Gross lesions are not present.

The histological lesions are specific and with a diagnostic value. A nonpurulent encephalomyelitis with marked perivascular clusters is present. The IEM virus is from the Picornaviridae family. The previous studies placed the virus in the Enterovirus genus, but based on the most recent studies, it is provisionally referred to the Hepatovirus genus. The virus is found in faeces of infected chickens and could survive there for at least 4 weeks. It is relatively highly resistant to environmental conditions.

A particularly valuable finding is the central chromatolysis of neurons in segments from the lumbosacral widening of the spinal cord and more rarely, chromatopyknosis. The infection occurs in non-vaccinated broiler breeder flocks and its course is subclinical. A vertical transmission of the infection to susceptible birds is realised by the eggs laid during this period. At the same time, parents acquire immunity and transmit it to the next generations thus protecting them from IEM. So, clinical manifestation of IEM is observed in chickens, hatched from the eggs during the 2- or 3 week period after breeder flock's infection. The next generations are immune.

Valuable diagnostic findings are the dense lymphoid clusters in the muscles of the proventriculus and the gizzard and in the pancreatic interstitium. The history of the disease, the age of onset and the typical nervous signs, especially the head tremor are indicative for the diagnosis that could be finally confirmed by histology. IEM should be differentiated from other diseases with nervous symptoms as encephalomalacia, mycotic encephalites, toxicoses (salt, pesticides). Prevention -vaccination of breeder flocks with regard to ensuring maximum protection of their offspring. The chickens from naturally infected flocks obtain an adequate maternal immunity that preserves them from the disease.

Newcastle Disease

The Newcastle disease (ND) is a highly contagious disease in many species of domestic, exotic and wild birds that, depending on its tropism, is characterised by marked variations in morbidity, death rate, symptoms and lesions. The clinico-morphological signs possess a distinct viscerotropic or neurotropic character. In the viscerotropic form, haemorrhagic diphtheritic lesions of the entire alimentary tract, from the beak to the vent, are present. The haemorrhages of gizzard epithelium are remarkable. The mucous coat is oedematous, covered with thick mucus and mottled with haemorrhages varying from single to multiple, sometimes gathered at the boundaries with the gizzard or the oesophagus.

Typical for this form are the haemorrhagic necrotic and focal diphtheroid lesions affecting the mucosa of the buccal cavity, the stomach and the intestines. The disease is generally prevalent in hens, more rarely in turkeys, exotic or wild birds. Birds at any age are susceptible. It is caused by a paramyxovirus. Depending on their pathogenicity, the numerous known strains are classified as lentogenic, meso-genic and velogenic. The vaccines made of lentogenic strains provoke a shorter immunity that requires a revaccination. The vaccines from mesogenic strains result in a lasting immunity, but could provoke a lethal issue especially in birds without a primary immunity created on the basis of lentogenic vaccinal strains.

A frequent finding is the enlargement and haemorrhages of caecal tonsils and haemorrhagic cloacitis. Usually, these lesions begin from the lymphoid tissue of the mucous coat. Virus-containing excreta of infected birds, that contaminate the forage, water and the environment, are the source of infection.

The infection is transmitted mainly by an oral route, the airborne or contact transmission being more infrequent. The virus, contained in incubated eggs, results in embryo's death and then perishes. There is no permanent carriership of the virus. An important factor in the transmission of velogenic viruses could be the exotic birds and fighting cocks. The death rate could arrive at 70-100%.

The neurotropic form of the disease is clinically manifested by ataxia, opisthotonus, torticolis, paresis and paralysis of legs. This form is frequently accompanied by respiratory symptoms. Histo-pathologically, the picture of non-purulent lymphocytic encephalomyelitis is observed. The lesions of paramyxovirosis in pigeons are entirely identical. On the basis of the history and the clinico-morphological signs, a tentative diagnosis could be made, but its laboratory confirmation is mandatory. ND should be distinguished by avian influenza, fowl cholera, etc.

Fowl Pox

Fowl pox (FP) is a viral disease in hens, turkeys and many other birds, characterised by cutaneous lesions on the feather-less skin and/ or diphtheritic lesions of mucous coats of the upper alimentary and respiratory tract. FP is encountered in either cutaneous or diphtheritic form or in both. In most outbreaks, the cutaneous form is prevailing. The lesions vary according to the stage of development: papules, vesicles, pustules or crusts. The lesions are usually in the region of the head. Lesions around the vent in a pigeon. FP is caused by an epitheliotropic DNA virus from the Avipox genus, the Poxviridae family. Some virus types (strains) exist: fowl pox virus, turkey pox virus, pigeon pox virus, canary pox virus, etc., different in pathogenic and immunogenic aspects.

The viruses are very resistant to environmental factors and could persist for several months.

Frequently, the conjunctival mucosa, injured by the pox virus, is an entrance door for additional contamination (£. coli, Staphylococcus spp., etc.) and development of complications. The infection is mechanically spread by dissemination of the virus through desquamation of crusts that contain it. Some mosquitoes and blood suckling arthropods could also distribute the virus. The mosquitoes remain infective for several weeks. The incubation period is from 4 to 10 days. The disease is spread slowly and many weeks could pass between its emergence and severe outbreaks occurrence.

Diphtheritic lesions look like whitish or yellowish plaques that are deposited and grown on the mucous coats of the buccal and nasal cavities, the sinuses, the larynx, the pharynx, the trachea or the oesophagus (arrows). The diagnosis is made upon the typical cutaneous and diphtheritic lesions. The prevention is performed by vaccinations that could be made at any age, if necessary.

Reovirus Infections

Viral Arthritis/tenosynovitis: The most frequent reovirus-associated disease in poultry is viral arthritis. Clinically it is manifested by lameness and swellings affecting primarily tarsometatarsal joints and the feet. Many infected birds are in a good general condition, but some could be lethargic and exhausted. The mortality is very low. The infections affect predominantly meat type poultry.

In some cases, joint cavities or tendon sheaths contain a small amount of straw-yellow exudate whereas in other the exudate is haemorrhagic or fibrinous.

Fibrinous Tenosynovitis: The inflammation of the tendon progresses to a chronic type lesion characterised by tissue fibrosis in the affected area. The aetiological agent is a reovirus, member of the Orthoreovirus genus, Reoviridae family. Several serotypes are identified. For their identification, the agar gel precipitation test could be used. Reoviruses are highly resistant to a number of environmental factors (temperature, pH, etc.).

Stunting Syndrome in Broilers: The stunting syndrome in broilers is associated with a reovirus infection but according to some studies, the role of the reovirus is probably secondary. It is characterised by a considerably reduced live weight in affected birds and a various degree of ununiformity in the flock varying from 5-10% to 40-50%, usually seen after the age of 14 days.

The growing primary wing feathers are abnormally big for chickens with retarded growth, they protrude at various angles, so the disease is termed helicopter disease". One-day old chickens are the most susceptible to the infection.

Usually, a High-degree Atrophy of the Pancreas is Observed: Reoviruses are shed with faeces and could contaminate the egg shells. The transmission of the infection to susceptible chickens is realised horizontally. The vertical route of transmission is also proved. Reoviruses could persist in infected birds for more than 40 weeks.

The Small Intestine is Pale, Dilated and Contain Indigested Forage: The tentative diagnosis of viral arthritis and stunting syndrome could be made on the basis of symptoms and lesions. The detection of reovirus antibodies via ELISA supports the diagnosis. Viral arthritis should be differentiated from M. synov/ae-induced, staphylococcal arthrites and the spontaneous rupture of the tendon of the gastrocnemius muscle. The vaccination of broiler breeder flocks with live or inactivated vaccines protects one-day-old chicks. Infected birds should be removed from the premises. The iodine disinfectants are considered effective for inactivation of agents.

Chapter 9

Avian Infectious Laryngotracheitis

Introduction

Gallid Herpes virus causes respiratory disease in chickens and pheasants.

Disease varies from mild to peracute, with mortality in peracute outbreaks exceeding 50%. As with all herpesviruses, GHV-1 can remain latent in carriers after infection and then be shed intermittently, recrudescing with stress.

Signalment

The chicken is the primary host and reservoir host. A form of LT has been described in pheasants.

Distribution

Worldwide. Transmission is via direct contact and contaminated people and equipment. Vermin and wild birds and dogs may aid mechanical transmission.

Clinical Signs

Respiratory signs:

- Nasal discharge which is often bloody
- Coughing which may also include blood
- Sneezing, dyspnoea, gasping, upper respiratory tract pain
- Abnormal lung sounds
- Decreased egg production, thin egg shells, lack of growth
- Neurological and ophthalmologic signs may develop.

Death may occur rapidly and with high mortality in peracute and acute disease. In recent times, LT usually presents in a mild form and most birds recover.

Diagnosis

On post-mortem, haemorrhagic tracheitis and bloodstained mucus are evident. Pneumonia and sacculitis may also be seen. Caseous diptheritic membranes may be present on the mucosae of the upper respiratory tract. Histopathology reveals loss of cilia, mucosal gland atrophy, intranuclear inclusion bodies and epithelial cell sloughing. Characteristic syncytia develop. A fibrinonecrotic membrane may be present in more chronic disease cases.

Antigen ELISA is both straightforward, quick and sensitive. The PCR can be used to detect LTV. Immunofluorescent or Immunoperoxidase staining can also be performed and is more rapid but less sensitive. Virus isolation on a variety of tissues including tracheal swabs or tissue samples may be useful.

Agar Gel Immunodiffusion can detect virus in tracheal samples.

Electron microscopy can be used to demonstrate viral particles in tracheal scrapings or exudates but is insensitive. Measuring viral antibody measures infection indirectly as serum antibodies peak around 2 weeks after infection and wane slowly afterwards.

Treatment

Where early diagnosis is made, vaccination can be administered in the face of infection to help reduce further morbidity and mortality.

Control

ILT can be effectively controlled by vaccination. Vaccinated and unvaccinated birds should not be mixed due to the possibility of reversion to virulence. Most are modified live isolates and are administered by eye drop.

Adequate Biosecurity, Quarantine and Disinfection is also Essential. Wild birds and vermin should be prevented from accessing poultry and their food/water sources.

Gapeworm

A gapeworm (*Syngamus trachea*) is a parasitic nematode worm infecting the tracheas of certain birds. The resulting disease, known as gape or the gapes, occurs when the worms clog and obstruct the airway. The worms are also known as red worms or forked worms due to their red colour and the permanent procreative conjunction of males and females. Gapeworm is common in young, domesticated chickens and turkeys.

When the female gapeworm lays her eggs in the trachea of an infected bird, the eggs are coughed up, swallowed, then defecated. When birds consume the eggs found in the feces or an intermediate host such as earthworms, snails (*Planorbarius corneus, Bithynia tentaculata, ...*), or slugs, they become infected with the parasite.

Ivermectin is a drug often used to control gapeworm infection in birds.

Morphology

Males and females are joined together in a state of permanent copulation forming a Y shape "forked worm". They are also known as the "red worm" because of their colour. Females are much larger (up to 20 mm long) than males (up to 6 mm long). The life history of the gapeworm is peculiar in that transmission from bird to bird may be successfully accomplished either directly (by the feeding of embryonated eggs or infective larvae) or indirectly (by ingestion of earthworms containing free or encysted gapeworm larvae they had obtained by feeding on contaminated soil).

Life Cycle and Pathogenesis

The life cycle is complicated in both its preparasitic and parasitic phases. In the preparasitic phase, L3s develop inside the eggs at which time they may hatch. Earthworms play an important role in the life cycle, serving as transport (paratenic) hosts. Larvae have been shown to remain viable for more than three years encapsulated in earthworm muscles. Other invertebrates may also serve as paratenic hosts and these include terrestrial snails and slugs as well as the larvae of Musca domestica (the common house fly) and Lucilia sericata (the green bottle fly responsible for cutaneous myiasis).

The parasitic phase involves substantial migration in the definitive host to reach the predilection site. Young birds are most severely affected with migration of larvae and adults through the lungs causing a severe pneumonia. Lymphoid nodules form at the point of attachment of the worms in the bronchi and trachea. Adult worms also appear to feed on blood. Worms in the bronchi and trachea provoke a hemorrhagic tracheitis and bronchitis with formation of large quantities of mucus, plugging the air passages and, in severe cases, causing asphyxiation. Pheasants appear to be particularly susceptible to infections resulting in mortality rates as high as 25% during outbreaks. The rapidly growing worms soon obstruct the lumen of the trachea and cause suffocation. Turkey poults, baby chicks and pheasant chicks are most susceptible to infection. Turkey poults usually develop gapeworm signs earlier and begin to die sooner after infection than young chickens. Lesions are usually found in the trachea of turkeys and pheasants but seldom if ever in the tracheas of young chickens and guinea fowl. The male worm, in the form of lesions, remains permanently attached to the tracheal wall throughout the duration of its life. The female worms apparently detach and reattach from time to time in order to obtain a more abundant supply of food.

Epidemiology

Earthworm transport hosts are important factors in the transmission of Syngamus trachea where poultry and game birds are reared on soil. The longevity of L3s in earthworms (up to 3 years) is particularly important in perpetuating the infection from year to year. Wild birds may serve as reservoirs of infection and have been implicated as the sources of infections in outbreaks on game-bird farms as well as poultry farms. Wild reservoir hosts may include pheasants, ruffed grouse, partridges, turkeys, magpies, meadowlarks, robins, grackles, jays, jackdaws, rooks, starlings and crows.

There is also evidence to suggest that strains of Syngamus trachea from wild bird reservoir hosts may be more infective for domestic birds if they first pass through an earthworm transport host rather than by direct infections via ingestion of L3s or eggs containing L3s. Clinical signs Blockage of the bronchi and trachea with worms and mucus will cause infected birds to gasp for air.

They stretch out their necks, open their mouths and gasp for air producing a hissing noise as they do so. This "gaping" posture has given rise to the common term "gapeworm" to describe Syngamus trachea. These clinical signs first appear approximately 1-2 weeks after infection. Birds infected with gapeworms show signs of weakness and emaciation and usually spend much of their time with eyes closed and head drawn back against the body. An infected bird may give its head a convulsive shake in an attempt to remove the obstruction from the trachea so that normal breathing may be resumed. Severely affected birds, particularly young ones, will deteriorate rapidly; they stop drinking and become anorexic. At this stage, death is the usual outcome. Adult birds are usually less severely affected and may only show an occasional cough or even no obvious clinical signs.

Diagnosis

A diagnosis is usually made on the basis of the classical clinical signs of "gaping". Subclinical infections with few worms may be confirmed at necropsy by finding copulating worms in the trachea and also by finding the characteristic eggs in the feces of infected birds. Examination of the trachea of infection birds shows that the mucous membrane is extensively irritated and inflamed. Coughing is apparently the result of this irritation to the mucous lining.

Control and Treatment of Syngamus Trachea

Prevention

In the artificial rearing of pheasants, gapes are a serious menace. Confinement rearing of young birds has reduced the problem in chickens compared to a few years ago. However, this parasite continues to present an occasional problem with turkeys raised on range. Confinement rearing of broilers/pullets and caging of laying hens, have significantly influenced the quantity and variety of nematode infections in poultry. For most nematodes, control measures consist of sanitation and breaking the life cycle rather than chemotherapy. Confinement rearing on litter largely prevents infections with nematodes using intermediate hosts such as earthworms or grasshoppers, which are not normally found in poultry houses. Conversely, nematodes with direct life cycles or those that utilise intermediate hosts such as beetles, which are common in poultry houses, may prosper.

Treatment of the soil or litter to kill intermediate hosts may be beneficial. Insecticides suitable for litter treatment include carbaryl, tetrachlorvinphos (stirofos). However, treatment is usually done only between grow-outs. Extreme care should be taken to ensure that feed and water are not contaminated. Treatment of range soil to kill ova is only partially successful. Changing litter can reduce infections, but treating floors with oil is not very effective. Raising different species or different ages of birds together or in close proximity is a dangerous procedure as regards parasitism. Adult turkeys, which are carriers of gapeworms, can transmit the disease to young chicks or pheasants, although older chickens are almost resistant to infection.

Treatment

Thiabendazole (Tresaderm) is currently approved for use only in pheasants and is effective when administered in the feed. Continuous medication of pen-reared birds has been recommended, but is not economical. Several other compounds have been shown effective against S. trachea under experimental conditions.

Methyl 5-benzoyl-2-benzimidazole was 100% efficacious when fed prophylactically to turkey poults. 5-isopropoxycarbonylamino-2-(4-thizolyl)-benzimidazole was found to be more efficacious than thiabendazole or disophenol.

The level of control with three treatments of cambendazole on days 3-4, 6-7, and 16-17 post-infection was 94.9% in chickens and 99.1% in turkeys. Levamisole (Ergamisol), fed at a level of 0.04% for 2 days or 2 g/gal drinking water for 1 day each month, has proven effective in game birds. Fenbendazole (Panacur) at 20 mg/kg for 3-4 days is also effective.

Infectious Bursal Disease

Etiology and Transmission

Infectious bursal disease is caused by a birnavirus (IBDV) that is most readily isolated from the bursa of Fabricius but may be isolated from other organs. It is shed in the feces and transferred from house to house by fomites. It is very stable and difficult to eradicate from premises.

IBDV may be isolated in 8 to 11 day old, antibody-free chicken embryos with inocula from birds in the early stages of disease. The chorioallantoic membrane is more sensitive to inoculation than is the allantoic sac. IBDV also may be isolated in cell cultures derived from the cloacal bursa and established cell lines, and some strains may be isolated in chicken-embryo fibroblasts.

Cell-culture-adapted strains of IBDV produce a cytopathic effect and may be used for quantitative serologic tests. Two serotypes of IBDV have been identified; within them, antigenic variation between strains is considerable. Serotype 2 infects chickens and turkeys but does not cause clinical disease or immunosuppression.

"Variant" strains of IBDV, which have major antigenic differences from the "standard" strains, cause immunosuppression but not clinical disease in older chickens.

Clinical Findings

Infectious bursal disease is highly contagious; results of infection depend on age and breed of chicken and virulence of the virus. Infections may be subclinical or clinical. Infections before 3 wk of age are usually subclinical. Chickens are most susceptible to clinical disease at 3-6 wk, but severe infections have occurred in Leghorn chickens up to 18 wk old. Early subclinical infections are the most important form of the disease because of economic losses. They cause severe, long-lasting immunosuppression due to destruction of immature lymphocytes in the bursa of Fabricius, thymus, and spleen. The humoral (B cell) immune response is most severely affected; the cell-mediated (T cell) immune response is affected to a lesser extent.

Chickens immunosuppressed by early IBDV infections do not respond well to vaccination and are predisposed to infections with normally nonpathogenic viruses and bacteria. Common diseases are usually exacerbated by IBDV infections. Subclinical infections by the "variant" strains occur in immature birds, and severe longterm immunosuppression and bursal atrophy result from early infections.

In clinical infections, onset of the disease is sudden after an incubation of 3-4 days. Chickens exhibit severe prostration, incoordination, watery diarrhea, soiled vent feathers, vent picking, and inflammation of the cloaca. Losses range to >20%. Recovery occurs in <1 wk, and broiler weight gain is delayed by 3-5 days. The presence of maternal antibody will modify the clinical course of the disease. Virulence of field strains of the virus varies considerably. Very virulent (vv) strains of the virus that cause high mortality and morbidity were detected first in Europe. These spread throughout the Old World in the last decade and in 1999 were in South America. The vv strains have not been detected in the USA.

Lesions

At necropsy, the cloacal bursa is swollen, edematous, yellowish, and occasionally hemorrhagic, especially in birds that have died of the disease. Congestion and hemorrhage of the pectoral, thigh, and leg muscles is common. Chickens recovered from IBDV infections have small, atrophied, cloacal bursas due to the destruction and lack of regeneration of the bursal follicles.

Control

There is no treatment. Depopulation and rigorous disinfection of contaminated farms have achieved limited success. Live vaccines of chick-embryo or cell-culture origin and of varying virulence can be administered by eye drop, drinking water, or SC routes at 1-21 days of age. The immune response can be altered by maternal antibody, and the more virulent vaccine strains can override higher levels of antibody.

High levels of maternal antibody during early brooding of chicks in broiler flocks (and in some commercial layer operations) can minimise early infection, subsequent immunosuppression, or both. Breeder flocks should be vaccinated one or more times during the growing period, first with a live vaccine and again just before egg production with an oil-adjuvanted, inactivated vaccine. Inactivated vaccines of chick-embryo, bursa, or cell-culture origin are available. The latter vaccines induce higher, more uniform, and more persistent levels of antibody than do live vaccines. The immune status of breeder flocks should be monitored periodically with a quantitative serologic test such as virus neutralisation or ELISA. If antibody levels fall, hens should be revaccinated to maintain adequate immunity in the progeny.

Bloodborne Organisms

Avian blood may contain various disease agents including viruses, bacteria, rickettsiae, protozoa, microfilariae, and rarely fungi. Except for viruses, these organisms often can be identified by microscopic examination of wet mounts, buffy coat, or blood smears; appropriate culturing techniques; or subinoculation of blood into susceptible birds. Microscopically, some are within blood cells (Plasmodium, Haemoproteus, Leucocytozoon, Atoxoplasma, Hepatocystis, Babesia, Aegyptianella), while others are free in the plasma (Trypanosoma, microfilariae, bacteria, spirochetes). None lives exclusively in the blood; most are found in tissues but are present in blood during part of their life cycle. Some, such as microfilariae and Plasmodium, may have a periodicity when numbers or stages of parasites are present at different times. In such cases, examining multiple smears at intervals will increase the likelihood of obtaining a diagnosis. Seasonal variations in infection rates relate to the activity of arthropod vectors. When possible, tissue cytology is also a useful adjunct to examination of blood. Several bloodborne organisms are either uncommon or not associated with clinical disease. However, weakened or injured raptors infected with hemoprotozoa had higher mortality and delayed recovery compared with uninfected birds. Routine examination for bloodborne organisms should be included in the clinical and diagnostic procedures for any ill bird.

Thin blood smears should be made with blood directly from the bird if possible. Anticoagulants, storage, and cooling of the blood can distort protozoan morphology and introduce artifacts. A small drop of blood can be collected using a syringe and needle, or by selecting a small vessel in the wing web and, after cleaning the site thoroughly with alcohol and letting it dry, puncturing the vessel with a lancet so that a small drop of blood wells up from the wound. The drop should be picked up without touching the skin and spread on a clean glass slide at a 30° angle to make a thin smear. A good quality Romanowsky-type stain that gives good polychromatic coloration (eg, Giemsa stain) should be used. At least 200 oil-immersion fields (~20,000 RBC) for single smears or 100 for multiple smears from the same bird should be examined. Leucocytozoon and microfilariae are found around the periphery of smears and can be easily seen on low-power magnification.

Bloodborne organisms in plasma or WBC are concentrated in the buffy coat. A diamond-tipped pencil can be used to cut the microhematocrit tube just below the buffy coat above the packed RBC. The buffy coat should be expressed from the cut end with a small amount of plasma to make a suspension, and a thin smear prepared. Stained buffy coat smears are recommended for detecting bacteremia, spirochetes, and chronic Leucocytozoon, Trypanosoma, or Atoxoplasma infections. An excellent technique for identifying low numbers of motile organisms such as spirochetes and microfilariae is direct examination of the buffy coat by darkfield or phase contrast microscopy. The buffy coat and all of the plasma should be expressed onto a glass slide, and covered with a coverglass, which is depressed slightly to spread the buffy coat. The buffy coat/plasma interface should be examined with darkfield or reduced light microscopy to detect motile organisms.

Plasmodium infections can be determined by subinoculation. Ideally, birds of the same or a known susceptible avian species should be used, but this is often not practical. In general, canaries are used for detecting passerine infections, and turkeys are susceptible to most plasmodia that infect gallinaceous birds. Parasites remain viable in blood stored at 32°F (in ice) for at least 7 days. Inoculation IV is preferred and will result in earlier parasitemia, but any parenteral route can be used. Recipients should be examined twice weekly for a minimum of 4 wk if exposed IV; longer times are needed if other routes of inoculation are used. Spirochete, Aegyptianella, and bacterial infections can also be detected by subinoculation of infectious blood; bacteria can usually be identified by blood culture.

To make a diagnosis of infection with an intracellular blood protozoan on a thin blood film, it first should be determined that the "parasites" in question are neither normal nor artifact. The following should then be determined: the host cell and whether it is normal or deformed beyond identification, whether pigment granules are present or absent, and whether merogony is occurring. Identification of an organism beyond genus (or subgenus in the case of Plasmodium spp) is difficult and usually unnecessary for clinical purposes.

Aegyptianellosis

Aegyptianellosis is an acute, tickborne, febrile disease caused by Aegyptianella spp, a rickettsia in the family Anaplasmataceae. Infection of avian species, including chickens, turkeys, guineafowl, quail, pigeons, crows, waterfowl, ratites, passerines, and psittacines has been described. Ticks, especially Argas spp, transmit the organism; infection can also be reproduced by blood inoculation. Organisms appear as single or multiple, round, "signet-ring" (0.5-4 μm) or irregular oval bodies in RBC often lateral to the nucleus. Infections are most common in tropical and subtropical areas of Africa, Asia, and Europe; infection of wild turkeys in Texas has also been reported.

In endemic areas, infection is mild or asymptomatic. Ruffled feathers, anorexia, droopiness, diarrhea, fever, jaundice, and high mortality in younger birds occur in introduced or otherwise susceptible birds. Anaemia, which can lead to right heart failure and ascites, enlargement of the liver and spleen, enlarged discoloured kidneys, and pinpoint serosal hemorrhages are seen. Infestation with larval argasid ticks and Borrelia infection (spirochetosis, *Avian Spirochetosis: Introduction*) may accompany the disease.

Tetracyclines, especially doxycycline, are effective in controlling the disease and eliminating the organism from chronically infected birds. Tick control is an important adjunct to treatment.

Atoxoplasmosis (Isosporosis, Lankesterellosis)

Atoxoplasma are pale-staining, nonpigmented, oval, intracytoplasmic bodies within mononuclear cells currently thought to be lymphocytes. Usually, cells contain a single parasite, but multiple organisms can be seen in severe, acute infections. Presence of the protozoan causes the nucleus to curve around it giving the appearance that the organism is located within an indentation of the nucleus. At least 2 different genera of coccidian protozoa (Isospora, Lankesterella) have merozoites in the lymphocytes that are indistinguishable from each other and have been called Atoxoplasma. Passerine birds, especially canaries, finches, sparrows, and species of the Sturnidae family (starlings, mynahs) are affected by both protozoa. Poultry are not known to be affected.

Isospora has a direct life cycle that includes an extra-intestinal, systemic phase in some species. Merogony occurs in the intestine and oocysts are passed in the feces. The systemic part of the life cycle was initially not recognised, leading to the description of the intestinal stages and oocysts as Isospora serini. Transmission is faecal-oral via ingestion of oocysts in droppings from infected birds. Infected canaries can shed oocysts for at least 2 yr. Most species of avian Isospora do not have extra-intestinal stages. In canaries, I serini has a systemic phase while I canaria infects only the intestinal tract.

Lankesterella has stages in the blood indistinguishable from those caused by systemic Isospora spp, but merogony occurs in tissues other than the intestine, especially the lungs. The life cycle is indirect. Bloodsucking arthropods, particularly mites, become infected during feeding and transmit the protozoan when it refeeds on a susceptible bird.

Signs include listlessness, diarrhea, and anorexia. Mortality can be high (up to 80%) in young birds. In acutely affected birds, there is marked hepatomegaly and splenomegaly, often with multifocal necrosis. The enlarged liver and gallbladder can be seen through the abdominal wall, especially if it is moistened with alcohol, which provides the basis for the common name black spot disease. High numbers of parasites infecting lymphocytes are present in blood and organ impression smears. Nearly spherical oocysts averaging 19×21 μm are present in droppings of canaries infected with Isospora. Oocysts of I serini need to be distinguished from those of I canaria, which are slightly larger and more oval.

Diagnosis is difficult in chronically infected older birds. Very few parasites are present in blood and tissues, and oocysts are shed intermittently, although sometimes in high numbers. Buffy coat and organ smears are preferred. Negative findings should not be interpreted to mean that infection is not present. Hepatic and splenic enlargement persists because of infiltrations with high numbers of large lymphoid cells that serve as host cells for the parasite. Histopathologically, organisms are difficult to find and identify. Lesions may be mistaken for lymphosarcoma.

There is no known effective treatment. Anticoccidial drugs do not affect parasites in the tissues. Use of antimalarial drugs has been suggested. Good management procedures, including prevention of exposure to mites, isolation of age groups, and scrupulous cleanliness (particularly daily cleaning before oocysts sporulate) help control the disease. Disinfectants have little effect on oocysts.

Filariasis

Microfilariae are commonly found in the blood of wild birds but are rare to absent in poultry except in southeast Asia where infections in chickens and waterfowl occur. A high percentage of imported cockatiels have microfilariae. At least 16 genera of filarids are found in avian species. All have an indirect life cycle with bloodsucking insects (eg, lice, mosquitos, midges) serving as intermediate hosts.

Adults are relatively short lived and mature in body cavities, including the eye and ventricles of the brain, respiratory system, cardiovascular system, or connective tissues; some produce characteristic subcutaneous nodules. In contrast, microfilariae are long lived and may be numerous in the skin as well as in the circulation. Increased numbers of microfilariae have been seen in stressed individuals, but they rarely cause clinical disease or mortality. A possible exception is infection of emus with Chandlerella, a common filarid of the brain of free-living grackles. Parasites apparently do not produce microfilariae iemus. Affected emus show signs of CNS disease. Treatment with ivermectin, levamisole, or injection of nodules with 0.5% potassium permanganate solution, and surgical removal of adults have been used.

Haemoproteus Infection

Infections with Haemoproteus spp are common in nondomestic birds. Pigeons and doves are frequently infected. Species are found in free-living ducks, quail, and turkeys but are rare to absent in commercial flocks probably because of very specific feeding habits of Culicoides spp and hippoboscid flies, the invertebrate vectors. In free-living bird populations, females are more frequently infected than males. Until recently, Haemoproteus was considered relatively innocuous and of little clinical significance. Fatal infections can occur because of extensive widespread necrosis accompanying development of large exoerythrocytic megalomeronts in muscle, heart, liver, and lung. Mortality as high as 78% has occurred in bobwhite quail infected with Isospora lophortyx. Similar meronts and lesions have been reported previously in cases of "aberrant leucocytozoonosis" and arthrocystosis.

The diagnostic presence of large, pigmented gametocytes in mature RBC that often partially or completely encircle the nucleus without merogony can follow or occur simultaneously with systemic involvement. Sudden death without clinical signs or a prolonged course of weakness, lameness, dyspnea, lethargy, poor growth, and anaemia may be seen. Little is known about effective treatment, although antimalarial drugs may be tried. Chloroquine (5 mg/kg) and buparvaquone (2.5 mg/kg) have been reported to be effective in treating pigeons; diminazene and quinapyramine were either ineffective or toxic. Measures to control invertebrate hosts should help prevent heavy infections.

Leucocytozoonosis

Infections with Leucocytozoon spp range from subclinical to fatal. Mortality may approach 100% but varies greatly with species and strain of parasite, host species, degree of exposure, and other factors. Acute outbreaks of leucocytozoonosis have been reported in chickens (Asia, Africa), turkeys (North America), waterfowl (North America, Europe), and a number of free-living and captive avian species throughout the world. Species in domestic birds include L simondi in waterfowl; L smithi in turkeys; and L caulleryi, L sabrazesi, L andrewsi, and L schoutedeni in chickens.

L caulleryi can be highly pathogenic, causing a lethal hemorrhagic disease of chickens in southeast Asia. Numerous Leucocytozoon spp infect nondomestic birds (eg, blood smears from raptors often contain gametocytes). Clinical disease and mortality result from anaemia caused by antierythrocytic factors produced by the parasite, high numbers of the large gametocytes blocking pulmonary capillaries, or parasites invading the endothelium of vessels in vital tissues (brain, heart, etc.) where they form megalomeronts that occlude vessels and result in multifocal necrosis.

Wild birds are reservoirs in some areas and are responsible for initiating infection in young birds each year. Parasitemia often increases dramatically in late April and early May (called spring rise), just before arthropod vectors, black flies (Simulium spp), or biting midges (Culicoides spp) increase. Ducks that have recovered from infection with L simondi relapse when light cycles are manipulated to increase egg production. Increased levels of prolactin have been suggested as a possible cause. Acute disease is seen more often in the young when they have high parasitemia and when black flies or biting midges are most abundant. Subacute or chronic disease is seen in the young outside fly season and in older birds at any season; parasitemia is usually low. Recovered birds remain carriers and serve as a reservoir for young, susceptible birds.

Clinical Findings, Lesions, and Diagnosis

Acutely affected birds are listless and have anaemia, leukocytosis, tachypnea, anorexia, diarrhea with green droppings, and often CNS signs. Egg production is impaired in laying chickens infected with L caulleryi. Signs are evident ~1 wk after infection and coincide with the onset of parasitemia. Visibly affected birds die after 7-10 days or may recover with sequelae of poor growth and egg production. Hemorrhages, splenomegaly, and hepatomegaly are seen. Grossly visible white dots in affected organs are megalomeronts.

In thin blood smears, gametocytes may be seen along the edges and tail of the smear. Leucocytozoon is identified by large gametocytes that lack pigment and distort the host cell (RBC or WBC), making it no longer identifiable. Shape of gametocytes varies with the species—some are elongated with long tapering extremities, while others are round. Serology may detect prior infection.

Treatment and Control

Treatment usually is not effective. Preventive medication with combined pyrimethamine (1 ppm) and sulfadimethoxine (10 ppm) combined in the feed controls L caulleryi; clopidol (0.0125-0.025%) controls L smithi. Measures to control invertebrate vectors are helpful. Humoral immunity resulting from vaccination will protect against L caulleryi infection.

Plasmodium Infection

Plasmodium spp, which often are not host-specific, infect a wide variety of domestic and wild birds in most areas of the world and can cause high losses. P gallinaceum infects chickens in Asia and Africa and causes low mortality in indigenous chickens but rates may be as high as 80-90% in commercial birds. P juxtanucleare infects chickens in Asia, Africa, and South America; most infections are mild or asymptomatic. P durae infects turkeys and gallinaceous birds other than chickens in Africa; mortality in turkeys can approach 100%. Clinical malaria has not been reported from poultry in North America, but indigenous wild turkeys can become infected with at least 4 different Plasmodium species. Asymptomatic infections in endemic or introduced birds can be spread via mosquitos and cause fatal disease in introduced (eg, zoo penguins) or resident (eg, Hawaiian avifauna) birds, respectively. Invertebrate hosts are ornithophilic mosquitos, usually Culex, Culiseta, or Aedes spp.

Clinical Findings, Lesions, and Diagnosis

Infection with Plasmodium spp may be nonclinical or cause illness characterised by weakness, lassitude, dyspnea, anaemia, abdominal distention, increased right heart weight, ocular hemorrhage, and death. Death results from severe anaemia or blockage of capillaries in the brain or other vital organs by exoerythrocytic meronts in endothelial cells. Liver and spleen are markedly enlarged and often discoloured (dark brown to black). Pigmented parasites including meronts are found in both immature and mture RBC. Infrequently, parasites are found in thrombocytes and WBC. In birds that die acutely, organisms may be sparse or absent in blood, but numerous meronts can be found in capillaries by examining squash or impression smears of brain, lung, liver, and spleen. Serologic and molecular diagnostic methods are under development but are not yet available for general use. Serology can detect infection when parasites are too few to be identified in blood smears.

Treatment and Control

Chemotherapy is variably effective in treating infected birds or flocks. Chloroquine (5-10 mg/kg) potentiated with primaquine (0.3 mg/kg), or chloroquine in drinking water (250 mg/120 mL) has been used. Grape or orange juice can disguise chloroquine's bitterness. Other antimalarial drugs have been used with success; quinacrine at 1.6 mg/kg/day for 5 days was successful for treating an infected peacock. Treatment with a combination of sulfamonomethoxine and sulfachloropyrazine is effective; halofuginone can be used for chemoprophylaxis in endemic areas. Persistent parasitemia or relapse may occur during and after treatment. Treatments should be evaluated on prevention of mortality and improvement of clinical disease. Birds that survive initial infection are generally refractory to subsequent infections. Prevention of exposure to mosquitos is a useful adjunct. Parasite development is reduced in mosquitos that survive previous infection with Bacillus thuringiensis *israelensis*.

Chapter 10

Epidemic Threat

Bird Flu is a Real Pandemic Threat to Humans

As their lungs filled ... the patients became short of breath and increasingly cyanotic. After gasping for several hours they became delirious and incontinent, and many died struggling to clear their airways of a blood-tinged froth that sometimes gushed from their nose and mouth. It was a dreadful business.

Isaac Starr, 3rd year medical student, University of Pennsylvania, 1918. All the genes of all influenza viruses in the world are being maintained in aquatic birds, and periodically they transmit to other species... The 1918 viruses are still being maintained in the bird reservoir. So even though these viruses are very ancient, they still have the capacity to evolve, to acquire new genes, new hosts. The potential is still there for the catastrophe of 1918 to happen again.

Dr. Robert Webster, Influenza Expert, Present Day.

The pair of quotes presented above were used to begin and end a short essay I wrote many years ago on The 1918 Spanish Flu Pandemic. Isaac Starr's words offer powerful testimonial to the harsh lethality of a killer stain of influenza, and Dr. Robert Websters observation reminds us that it would be dangerous to regard the events of 1918 as an isolated slice of history that we can safely file away and forget.

In the early part of 1995, as I cast around in search of the disease that would ultimately prove to be the true villian of the science thriller I was planning to write, influenza was scarcely on the radar scope of anyone but virologists specialising in the field of emerging diseases. Because influenza exists in the human population in an ever-present but relatively benign form, changing just a little from year to year, we tend, quite mistakenly, to think we understand the nature of the threat that it presents.

Bird Flu cannot be eradicated, we must learn to live with it.

Despite the apparent "novelty" of a strain of bird flu that can sweep across the globe and kill not only the domesticated foul that carries the virus causing the pandemic, but also those humans unfortunate enough to get too close to the sick birds they tend, finding influenza in domesticated birds is not the least bit uncommon.

The natural reservoir for influenza virus is in fact the intestines of water fowl. In particular, ducks habour the virus, and suffer no ill effects from doing so. Because these migratory birds are always on the move, and because they shed virus in the waterways that they inhabit, influenza virus travels widely and is easily transmitted from bird to bird.

Because of this, domesticated birds, like the turkey and the chicken, which cross paths with the virus, become infected and habor the potential to transmit the virus to humans (for example, virus secreted onto the surface of eggs could end up infecting egg handlers— for this reason, if bird handlers notice any of the following symptons in their birds they should immediately notify Federal or State animal health officials: sudden death; lack of energy and appetite; decreased egg production; soft-shelled or misshapen eggs; swelling; purple discoloration; nasal discharge; coughing, sneezing; lack of coordination and diarrhea).

Often an intermediate host, the pig, is involved in the cross-species transmission of influenza. When this happens, the result is referred to as Swine Flu. Even so, the source of the virus is generally aquatic birds.

Most strains of bird flu *do not* cross species and end up infecting humans. However the H5N1 strain that has in the last decade been known to infect humans, and kill them in 5 out of every 10 cases, clearly has surmounted the first of two barriers that must be cleared before a bird flu pandemic can take hold in the human population. The second barrier will be surmounted when this strain acquires the ability to jump from one human to another. This may never happen, and hopefully will not. But based on the history of pandemic flu, it seems more likely that this is just a matter of time. Whether this takes one year or twenty is just impossible to say. The World Health Organisation has estimated that a human-to-human strain of bird flu could result in an influenza pandemic with the potential to cause between 2 and 7 million fatalities worldwide. Considering the underwhelming state of health services offered by the majority of countries in the world today, these numbers do not appear to be overhyped estimates of the potential lethality of a bird flu pandemic. Should a bird flu pandemic break out with a lethality approaching that of the 1918 Spanish Flu pandemic, it is probably fair to state that the WHO prediction is likely to fall far short of the actual death toll.

Emergence of a Bird Flu Pandemic?

The current candidate for the cause of a bird flu pandemic that crosses into the human population is the H5N1 strain. As discussed in my essay on The 1918 Spanish Flu Pandemic, and the Emerging Bird Flu Pandemic, this strain was first detected in Hong Kong in 1997, and in the time since then has spread halfway across the globe despite strenuous efforts to eradicate it in poultry stocks that have become infected. This movement of the virus, which requires transmission from one (bird) host to another via shared waterways, is slow in comparison to the movement of an airborne strain which can be transmitted from one human host to another in the particulate matter exchanged during coughing and sneezing events. The pandemic of 1918 spread across the globe in about six months, despite the dependence of that time on ship borne travel. In an age of international travel, this encircling time could be reduced to a matter of days. Once it begins, the likelihood that it can be stopped in its tracks is small.

Life during a Bird Flu Pandemic

One can never predict with any accuracy the way a natural catastrophe will unfold. But some speculation on life during the time of a bird flu pandemic seems worthwhile. Things we can say for certain are the following: No country will be fully prepared for the event. This stems from the fact that no government is ever prepared for any natural catastrophe, as such preparations are prohibitively expensive. This means the bulk of the population will not have access to medical treatment when they need it. Families will be required to look after themselves. Only the patients identified in the very first stages of a bird flu pandemic are likely to receive treatment.

Medical supplies for combating a bird flu pandemic in the human population will be quickly exhausted. Even now, access to such antivirals as Tamiflu is poor in most parts of the world. (In late 2005 the makers of Tamiflu stepped up production to 10 times the normal level in response to worldwide demand for the drug in anticipation of an impending pandemic).

Likewise, flu vaccines, which require a long lead time for preparation (six months or more) and can easily be "spoiled" due to inadequate preparation, will not be available in anything like a sufficient supply in the foreseeable future.

Social disruption will be significant. When workers are faced with the decision of weighing the safety of family members against the worth of their job, you can bet the absentee rate will skyrocket during a pandemic. When the loss of workers begins to affect the delivery of food supplies, power, and water, we are suddenly looking at a different kind of emergency scenario—one reminiscent of the scenes that played out in New Orleans after Hurricane Katrina struck in 2005. When the services we take for granted every day suddenly disappear, people become desparate and things go to hell rather quickly. The break down of social order may be the most frightening prospect that an emerging bird flu pandemic could bring about.

Preparing for a Bird Flu Pandemic

It is tempting for all of us, the author included, to stick our heads in the sand and hope that the news stories about the possibly of an impending *bird flu pandemic* are mostly hype. But this is about as wise as living in Florida and hoping you have seen the last hurricane, or living in Los Angeles and assuming that you'll not be hit by another earthquake. Really it is just a matter of time, though no one knows how much.

There are some not so difficult things you can do to prepare, things that will also help prepare you for other emergency events. These seems practical to me:

Stock up on water: You can never have too much of this in storage, and it is cheap. Have at least a 3 month supply in stock at any time (6 months if practical), and rotate your supply so that you use up the oldest bottles at the rate you replace them with new ones. This way you will never have to worry about filtering existing water supplies if it comes to water shortages in your area.

Stock up on food supplies: This one is a little more difficult, but if you are lucky enough to have one of those huge freezers in your home, fill it up.

Rotate your food, so that it is never more than 3 (or 6) months old. Be sure to buy only the foods you normally buy, as you'll be eating them regardless of whether or not any disaster strikes.

Dry foods can be stored separately, but people tend too forget about them and only check their supplies in the advent of an emergency. So be prepared to toss these after a year when they are no longer edible.

Purchase some quantity of air filtration masks which, in the event of a true bird flu pandemic will be regarded as indispensible when you need to leave the house and mix with the general populace.

As regards what type of mask you should obtain if you are inclined to purchase a box or two (not a bad idea as you won't be able to do it if the time comes that you need to wear them) the Centres for Disease Control and Prevention advise health workers to use a "fit-tested respirator, at least as protective as a National Institute of Occupational Safety and Health (NIOSH)-approved N-95 filtering facepiece (i.e., disposable) respirator" as one of the possible precautions against airborne infection when working with patients with known or suspected SARS or Bird Flu. 3M manufactures such masks, and the latest variety with the "Cool Flow" exhalation valve for increased comfort during extended wearing periods seems to be the mask of choice.

For other possible preparedness items you might check out a Distributor of Flu Preparedness Items (naturally enough!).

If it comes down to it, and an infectious form of bird flu arrives in your country, one of the most important things you can do to avoid infection is practice safe hygiene. This means that you go into obsessive compulsive mode and assume that everything you touch has the potential to be carrying virus. It's somewhat pointless to assume this of the contents of your own home (unless you have a stricken person on site) but once you leave the house, it is a good operating assumption. This means under no circumstances do you touch your hands to your face while in public.

Children, and adults, who have the habit of putting their fingers in their eyes and noses are *extremely susceptible* to infection in such an environment.

When you return to your home, clothing goes into the laundry immediately and you wash your hands thoroughly with soap to rinse away any virus you might have come in contact with. You always use a face mask in public if you have access to one. Otherwise you make one. Common sense suggests that anything that inhibits the likelihood of inhaled aerosolised influenza is better than nothing. But a proper face mask is designed to be efficient at this task. These are designed to be used *once only* and discarded.

If you are a vendor, think about how you might change the operational nature of your business to stay afloat during a time of crisis. What can you do to reduce the likelihood that your employees will desert you? For instance, if you sell clothing you are very unlikely to receive my business if I must enter an enclosed space (a building) in order to purchase something. At such a time I do not want to breathe the air of a confined space. But if your stock was accessible in an open-air market I would probably be much more inclined. Employees will feel the same way, particularly if the goods in question are not critical to day to day survival. Finally, do some more reading on the subject of bird flu pandemic. The information about bird flu found on this page is by no means exhaustive. Reading resources are listed below in the Related Links section.

Could a Bird Flu Pandemic resemble the Spanish Flu Pandemic of 1918?

A lot has happened in the last decade as regards our scientific understanding of the genetic mechanisms responsible for the virulence of some strains of influenza. Much of this new understanding has been a direct result of the work of Dr. Jeffery Taubenberger and his team at the Armed Forces Institute of Pathology.

Back in 1998 I interviewed Dr. Taubenberger about his effort to reconstruct the genes of the 1918 Spanish Flu virus. At the time I had forseen the day when it would be possible to reconstruct the entire virus.

In October of 2005, after 10 years of work by Taubenberger's team, it was finally reported that the virus had been re-engineered. You can read about the announcement in the PBS article entitled 1918 Spanish Flu Offers Clues About Pandemic Viruses. Of particular interest is the observation that Taubenberger's work "published in the journal Nature in October 2005, offered some parallels between the Spanish flu virus and the H5N1 strain of the bird flu slowly spreading through Asia and recently turning up in other parts of the world." The implication here is clear. If the currently circulating H5N1 bird flu virus does acquire human-to-human transmission, the estimated mortality rates offered by the World Health Organisation for a bird flu pandemic could prove to be wildy optimistic.

Influenza Vaccine

The influenza vaccine, also known as flu shot, is an annual vaccine to protect against the highly variable influenza virus. Each injected seasonal influenza vaccine contains three influenza viruses: one influenza type A subtype H3N2 virus strain, one influenza type A subtype H1N1 (seasonal) virus strain, and one influenza type B virus strain.

Purpose and Benefits of Annual Flu Vaccination

"Influenza vaccination is the most effective method for preventing influenza virus infection and its potentially severe complications."

Deadly Epidemics each Winter

An influenza epidemic emerges during flu season each winter. There are two flu seasons annually, corresponding to the occurrence of winter in opposite months in the Northern and Southern Hemispheres.

Worldwide, seasonal influenza kills an estimated 250,000 to 500,000 people each year. Tens of thousands of Americans die in a typical flu season, but there are notable variations from year to year. In 2010 the Centres for Disease Control and Prevention (CDC) in the United States changed the way it reports the 30 year estimates for deaths from influenza. Now they are reported as a range from a low of about 3,300 deaths to a high of 49,000 per year over the past 30 years. The majority of deaths in the industrialised world occur in adults aged 65 and over. A review at the NIAID division of the NIH in 2008 concluded that "Seasonal influenza causes more than 200,000 hospitalisations and 41,000 deaths in the U.S. every year, and is the seventh leading cause of death in the U.S." The average total economic costs caused by the annual influenza outbreak in the U.S. have been estimated at over $80 billion.

The number of annual influenza-related hospitalisations is many times the number of deaths. "The high costs of hospitalising young children for influenza creates a significant economic burden in the United States, underscoring the importance of preventive flu shots for children and the people with whom they have regular contact..." In 2006 the United States began recommending influenza vaccinations for preschoolers but Canada did not follow suit until 2010, "thereby creating a natural experiment to evaluate the effect of the policy in the United States."

A Canadian study found emergency room visits significantly lower for 2- to 4 year olds in Boston than in Montreal through the period (34% fewer ER trips).

Vaccination of preschoolers may have reduced their likelihood of transmission of flu to older siblings and raised the chances that their parents would vaccinate older children as well, since there were also 18 percent fewer emergency room visits by 5- to 18 year olds in Boston than Montreal during the study period. In another six-year observational study, vaccination of children aged 6 months through 5 years was found to prevent illness in more than half.

National Advice on Flu Vaccination

In Canada, the National Advisory Committee on Immunisation, the group that advises the Public Health Agency of Canada, currently recommends that everyone aged 2 to 64 years be encouraged to receive annual influenza vaccination, and that children between the age of six and 24 months, and their household contacts, should be considered a high priority for the flu vaccine.

In the United States, "Routine influenza vaccination is recommended for all persons aged e"6 months." The sole group for whom routine influenza vaccination is still not recommended is infants less than six months of age. Within its blanket recommendation for general vaccination, the United States, the Centres for Disease Control and Prevention (CDC) emphasizes to clinicians the special urgency of vaccination for members of certain vulnerable groups, and their caregivers:

Vaccination is especially important for people at higher risk of serious influenza complications or people who live with or care for people at higher risk for serious complications.

Benefits of Vaccination

Vaccination against influenza is also, according to research published in July 2010, thought to be important for members of high-risk groups who would be likely to suffer complications from influenza, for example pregnant women and children and teenagers from six months to 18 years of age; In expanding the new upper age limit to 18 years, the aim is to reduce both the time children and parents lose from visits to pediatricians and missing school and the need for antibiotics for complications.

An added expected benefit would be indirect — to reduce the number of influenza cases among parents and other household members, and possibly spread to the general community.

Vaccination of school-age children has a strong protective effect on the adults and elderly with whom the children are in contact. Children born to mothers who received flu vaccination while pregnant are strongly protected from having to be hospitalised with the flu. "The effectiveness of influenza vaccine given to mothers during pregnancy in preventing hospitalisation among their infants, adjusted for potential confounders, was 91.5%."

Healthy, working adults who received influenza vaccine reported 25 percent fewer episodes of upper respiratory illness than those who received the placebo (105 vs. 140 episodes per 100 subjects, $P < 0.001$), 43 percent fewer days of sick leave from work due to upper respiratory illness (70 vs. 122 days per 100 subjects, $P = 0.001$), and 44 percent fewer visits to physicians' offices for upper respiratory illnesses (31 vs. 55 visits per 100 subjects, $P = 0.004$). The study, reported in the NEJM, estimated cost savings at $46.85 per person vaccinated, and concluded that "Vaccination against influenza has substantial health-related and economic benefits for healthy, working adults."

Influenza vaccination has been shown highly effective in health care workers, with minimal adverse effects. In a study of forty matched nursing homes, staff influenza vaccination rates were 69.9% in the vaccination arm versus 31.8% in the control arm. The vaccinated staff experienced a 42% reduction in sick leave from work (P=.03). A review of eighteen studies likewise found a strong net benefit to health care workers. The two of these eighteen studies that assessed the relationship of patient mortality relative to staff influenza vaccine uptake both found that higher rates of health care worker vaccination correlated with reduced patient deaths. An analysis of data and patient population health in New Mexico's 75 long-term care facilities nursing homes found that as vaccination rates of health care personnel with direct patient contact rose from 51 to 75 percent, the chances of a flu outbreak among patients in that facility went down by 87 percent. The New Mexico study showed that vaccinating health care personnel provided more protection to residents than vaccinating residents themselves.

In a 2010 survey of healthcare workers, 63.5% reported that they received the flu vaccine during the 2010-11 season, an increase from 61.9% reported the previous season. Health professionals with direct patient contact had higher vaccination uptake, such as physicians and dentists (84.2%) and nurse practitioners (82.6%).

Cross-protection

Annual seasonal flu vaccination may provide some level of protection against novel flu viruses. A number of studies suggest that seasonal flu vaccine may offer cross-protection, both against the H5N1-type (avian influenza) H5N1 infection and the 2009 flu pandemic (the H1N1 "swine flu.") Vaccine protection can be long-lasting. Participants who received vaccination against the swine flu in 1976 still enjoyed benefits 33 years later, exhibiting a significantly enhanced immune response to the 2009 pandemic H1N1.

History of the Flu Vaccine

Vaccines are used in both humans and nonhumans. Human vaccine is meant unless specifically identified as a veterinary, poultry or livestock vaccine.

Influenza

The first influenza pandemic was recorded in 1580; since this time, various methods have been employed to eradicate its cause. The etiological cause of influenza, the orthomyxoviridae was finally discovered by the Medical Research Council (MRC) of the United Kingdom in 1933.

Known Flu Pandemics

- 1889–90 — Asiatic (Russian) Flu, mortality rate said to be 0.75–1 death per 1000 possibly H2N2.
- 1900 — Possibly H3N8.
- 1918–20 – Spanish Flu, 500 million ill, at least 20–40 million died of H1N1.
- 1957–58 – Asian Flu, 1 to 1.5 million died of H2N2.
- 1968–69 – Hong Kong Flu, 3/4 to 1 million died of H3N2.
- 2009 - Swine Flu, caused by H1N1/09, 14,286 died.

Flu Vaccine Origins and Development

In the world wide Spanish flu pandemic of 1918, "Physicians tried everything they knew, everything they had ever heard of, from the ancient art of bleeding patients, to administering oxygen, to developing new vaccines and sera (chiefly against what we now call *Hemophilus influenzae*—a name derived from the fact that it was originally considered the etiological agent—and several types of pneumococci). Only one therapeutic measure, transfusing blood from recovered patients to new victims, showed any hint of success."

In 1931, viral growth in embryonated hens' eggs was reported by Ernest William Goodpasture and colleagues at Vanderbilt University. The work was extended to growth of influenza virus by several workers, including Thomas Frances, Wilson Smith and Macfarlane Burnet, leading to the first experimental influenza vaccines.

In the 1940s, the US military developed the first approved inactivated vaccines for influenza, which were used in the Second World War. Greater advances were made in vaccinology and immunology, and vaccines became safer and mass-produced. Today, thanks to the advances of molecular technology, we are on the verge of making influenza vaccines through the genetic manipulation of influenza genes.

Flu Vaccine Acceptance

According to the CDC: "Influenza vaccination is the primary method for preventing influenza and its severe complications. Vaccination is associated with reductions in influenza-related respiratory illness and physician visits among all age groups, hospitalisation and death among persons at high risk, otitis media among children, and work absenteeism among adults. Although influenza vaccination levels increased substantially during the 1990s, further improvements in vaccine coverage levels are needed".

The current egg-based technology for producing influenza vaccine was created in the 1950s. In the U.S. swine flu scare of 1976, President Gerald Ford was confronted with a potential swine flu pandemic. The vaccination program was rushed, yet plagued by delays and public relations problems. Meanwhile, maximum military containment efforts succeeded unexpectedly in confining the new strain to the single army base where it had originated. On that base a number of soldiers fell severely ill, but only one died. The program was cancelled, after about 24% of the population had received vaccinations. An excess in deaths of twenty-five over normal annual levels as well as 400 excess hospitalisations, both from Guillain-Barré syndrome, were estimated to have occurred from the vaccination program itself, illustrating that vaccine itself is not free of risks.

The result has been cited to stoke lingering doubts about vaccination. In the end, however, even the maligned 1976 vaccine may have saved lives. A 2010 study found a significantly enhanced immune response against the 2009 pandemic H1N1 in study participants who had received vaccination against the swine flu in 1976.

Current Status

Influenza research includes molecular virology, molecular evolution, pathogenesis, host immune responses, genomics, and epidemiology. These help in developing influenza countermeasures such as vaccines, therapies and diagnostic tools.

Improved influenza countermeasures require basic research on how viruses enter cells, replicate, mutate, evolve into new strains and induce an immune response. The Influenza Genome Sequencing Project is creating a library of influenza sequences that will help us understand what makes one strain more lethal than another, what genetic determinants most affect immunogenicity, and how the virus evolves over time. Solutions to limitations in current vaccine methods are being researched.

The rapid development, production, and distribution of pandemic influenza vaccines could potentially save millions of lives during an influenza pandemic. Due to the short time frame between identification of a pandemic strain and need for vaccination, researchers are looking at novel technologies for vaccine production that could provide better "real-time" access and be produced more affordably, thereby increasing access for people living in low- and moderate-income countries, where an influenza pandemic may likely originate, such as live attenuated (egg-based or cell-based) technology and recombinant technologies (proteins and virus-like particles).

As of July 2009, more than 70 known clinical trials have been completed or are ongoing for pandemic influenza vaccines. In September 2009, the US Food and Drug Administration approved four vaccines against the 2009 H1N1 influenza virus (the current pandemic strain), and expected the initial vaccine lots to be available within the following month.

Prospects for Universal Flu Vaccines

Many groups world wide are working on a universal flu vaccine that will not need changing each year, as the sector has been viewed as "increasingly hot". Companies pursuing the vaccine as of 2009 and 2010 include BiondVax, Theraclone, Dynavax Technologies Corporation, VaxInnate, Crucell NV, Inovio Pharmaceuticals, and Immune Targeting Systems (ITS).

In 2008 Acambis announced work on a universal flu vaccine (ACAM-FLU-ATM) based on the less variable M2 protein component of the flu virus shell.

In 2009, the Wistar Institute received a patent for using "a variety of peptides" in a flu vaccine, and announced it was seeking a corporate partner.

In 2010, the National Institute of Allergy and Infectious Diseases (NIAID) of the U.S. NIH announced a breakthrough; the effort targets the stem, which mutates less often than the head of the virus.

DNA vaccines such as VGX-3400X (aimed at multiple H5N1 strains) contain DNA fragments (plasmids). Inovios SynCon DNA vaccines include H5N1 and H1N1 subtypes.

In July 2011, F16 researchers created an antibody, which targets a protein found on the surface of all influenza A viruses called haemagglutinin.

F16 is the only known antibody that treats all 16 subtypes of the influenza A virus and might be the lynchpin for a universal influenza vaccine.

Other Vaccines are Polypeptide Based

Some universal flu vaccines have started early stage clinical trials.

- BiondVax are targeting the less variable stalk of the haemagglutinin molecule with Multimeric-001. This is aimed at type A (inc

H1N1) and Type B influenza and has started a phase IIa study.

- Dynavax have developed a vaccine N8295 based on two highly conserved antigens NP and M2e and their TLR9 agonist, and started clinical trials in June 2010.
- ITS's fp01 includes 6 peptide antigens to highly conserved segments of the PA, PB1, PB2, NP & M1 proteins, and has started phase I trials.

Based on the results of animal studies, a universal flu vaccine may use a two-step vaccination strategy — priming with a DNA-based HA vaccine followed by a second dose with an inactivated, attenuated, or adenovirus-vector–based vaccine.

Some people given a 2009 H1N1 flu vaccine have developed broadly protective antibodies which has raised hopes for a universal flu vaccine.

Clinical Trials of Vaccines

A vaccine is assessed by the reduction of the risk of disease that is produced by vaccination, the vaccine's *efficacy*. In contrast, in the field, the *effectiveness* of a vaccine is the practical reduction in risk for an individual when they are vaccinated under real-world conditions.

Measuring efficacy of influenza vaccines is relatively simple, as the immune response produced by the vaccine can be assessed in animal models, or the amount of antibody produced in vaccinated people can be measured, or most rigorously, by immunising adult volunteers and then challenging with virulent influenza virus. In studies such as these, influenza vaccines showed high efficacy and produced a protective immune response. For ethical reasons, such challenge studies cannot be performed in the population most at risk from influenza – the elderly and young children. However, studies on the effectiveness of flu vaccines in the real world are uniquely difficult. The vaccine may not be matched to the viruses in circulation that year; virus prevalence varies widely between years, and influenza is often confused with other influenza-like illnesses.

Nevertheless, multiple clinical trials of both live and inactivated influenza vaccines against seasonal influenza have been performed and their results pooled and analysed in several recent meta-analyses. Studies on live vaccines have very limited data, but these preparations may be more effective than inactivated vaccines. The meta-analyses examined the efficacy and effectiveness of inactivated vaccines against seasonal influenza in adults, children, and the elderly. In adults, vaccines show high efficacy against the targeted strains, but low effectiveness overall, so the benefits of vaccination are small, with a one-quarter reduction in risk of contracting influenza but no significant effect on the rate of hospitalisation.

However, the risk of serious complications from influenza is small in adults, so unless the effect from vaccination is large it might not have been detected. In children, vaccines again showed high efficacy, but low effectiveness in preventing "flu-like illness". In children under two the data are extremely limited, but vaccination appeared to confer no measurable benefit.

In the elderly, vaccination does not reduce the frequency of influenza, but seems to reduce pneumonia, hospital admission and deaths from influenza or pneumonia. However, the current data on the effectiveness of influenza vaccines in the elderly may be unreliable, due to high levels of selection bias.

Overall, the benefit of influenza vaccination is clear in the elderly and vaccination of children may be beneficial. Routine vaccination of adults is not predicted to produce significant improvements in public health. The apparent contradiction between vaccines with high efficacy, but low effectiveness, may reflect the difficulty in diagnosing influenza under clinical conditions and the large number of strains circulating in the population. In contrast, during an influenza pandemic, where a single strain of virus is responsible for illnesses, an effective vaccine could produce a large decrease in the number of cases and be highly effective in controlling an epidemic. However, such a vaccine would have to be produced and distributed rapidly to have maximum effect.

Effectiveness of Vaccine

The CDC reports that studies demonstrate that vaccination is a cost-effective counter-measure to seasonal outbreaks of influenza. However it is not perfect. A study led by Dr. David K. Shay in February, 2008 reported that:

> *"full immunisation against flu provided about a 75 percent effectiveness rate in preventing hospitalisations from influenza complications in the 2005-6 and 2006-7 influenza seasons."*

Influenza vaccine has been demonstrated to prevent disease and death, both in numerous controlled studies and in painstaking scientific reviews of these studies. However the rigor of the science of these studies has also been criticized. A 2006 Cochrane review of influenza vaccination in the elderly stated "The apparent high effectiveness of the vaccines in preventing death from all causes may reflect a baseline imbalance in health status and other systematic differences in the two groups of participants. In one observational study, a sharply lower risk of death or hospitalisation for pneumonia was seen in vaccinated persons:

Results: The relative risk of death for vaccinated persons compared with unvaccinated persons was 0.39 [95% confidence interval (95% CI), 0.33-0.47] before influenza season, 0.56 (0.52-0.61) during influenza season, and 0.74 (0.67-0.80) after influenza season. The relative risk of pneumonia hospitalisation was 0.72 (0.59-0.89) before, 0.82 (0.75-0.89) during, and 0.95 (0.85-1.07) after influenza season. Adjustment for diagnosis code variables resulted in estimates that were further from the null, in all time periods. A better vitality and lower pneumonia hospitalisation was thus observed in the vaccinated group in this non-randomised population study before, during, and after influenza season. This could, however, have been ascribed in whole or in part to self-selection bias, since vaccinated persons were already healthier before flu season. Some sub-populations have been assumed to benefit from vaccination in the absence of directly specific studies. For example, a 2008 Cochrane review of healthy children found "Influenza vaccines are efficacious in children older than two but little evidence is available for children under two.".

The CDC in 2010, after a review of extant studies, extended its guidelines to recommend that every child over 6 months be given the influenza vaccine. Vaccines were shown effective against the influenza strains they are designed to protect against, but this translated to only a modest impact on working days lost due to influenza-like infections in a 2007 Cochrane review on influenza vaccines in healthy adults. While a 2010 Cochrane review noted that "Influenza vaccines have a modest effect in reducing influenza symptoms and working days lost" it stated found no evidence of prevention of complications, such as pneumonia, or transmission.

The group most vulnerable to non-pandemic flu, the elderly, is also the least benefitted by the vaccine, with an average efficacy rate ranging from 40-50% at age 65, and only 15-30% past age 70. There are multiple reasons behind this steep decline in vaccine efficacy, the most common of which are the declining immunological function and frailty associated with advanced age. An influenza vaccine with four times the usual amount of antigen (Fluzone High Dose) has shown increased immune response in the elderly and has now been approved by the U.S. Food and Drug Administration (FDA). In a non-pandemic year, a person in the United States aged 50–64 is nearly ten times more likely to die an influenza-associated death than a younger person, and a person over age 65 is over ten times more likely to die an influenza-associated death than the 50–64 age group. Vaccination of those over age 65 reduces influenza-associated death by about 50%.

However, it is unlikely that the vaccine completely explains the results since elderly people who get vaccinated are probably more healthy and health-conscious than those who do not. Elderly participants randomised to a high-dose group (60 micrograms) had antibody levels 44 to 79 percent higher than did those who received the normal dose of vaccine. Elderly volunteers receiving the higher dose were more likely to achieve protective levels of antibody. As mortality is also high among infants who contract influenza, the household contacts and caregivers of infants should be vaccinated to reduce the risk of passing an influenza infection to the infant. Data from the years when Japan required annual flu vaccinations for school-aged children indicate that vaccinating children the group most likely to catch and spread the disease—has a strikingly positive effect on reducing mortality among older people: one life saved for every 420 children who received the flu vaccine. This may be due to herd immunity or to direct causes, such as individual older people not being exposed to influenza. For example, retired grandparents often risk infection by caring for their sick grandchildren in households where the parents can't take time off work or are sick themselves. In most years (16 of the 19 years before 2007), the flu vaccine strains have been a good match for the circulating strains. In other flu seasons like that of 2007/2008, the match was less useful. But even a mismatched vaccine can often provide some protection:

Antibodies made in response to vaccination with one strain of influenza viruses can provide protection against different, but related strains. A less than ideal match may result in reduced vaccine effectiveness against the variant viruses, but it still can provide enough protection to prevent or lessen illness severity and prevent flu-related complications. In addition, it's important to remember that the influenza vaccine contains three virus strains so the vaccine can also protect against the other two viruses. For these reasons, even during seasons when there is a less than ideal match, CDC continues to recommend influenza vaccination. This is particularly important for people at high risk for serious flu complications and their close contacts.

Comparing Flu Shot to Nasal Spray

Flu vaccines are available either as:

TIV (flu shot (injection) of trivalent (three strains; usually A/H1N1, A/H3N2, and B) inactivated (killed) vaccine) or

LAIV (nasal spray (mist) of live attenuated influenza vaccine.)

TIV works by putting into the bloodstream those parts of three strains of flu virus that the body uses to create antibodies; while LAIV works by inoculating against those same three strains that have been genetically modified to minimise symptoms of illness. LAIV is not recommended for individuals under age 2 or over age 50, but might be comparatively more effective among children over age 2.

A military study on military personnel showed that flu shots yielded less illness than nasal spray. Based on one of the largest head-to-head studies comparing LAIV and TIV (which was conducted by the U.S. Armed Forces Surveillance Centre on military personnel who were stationed in the United States during three flu seasons from 2004 through 2007), investigators concluded that: "It may be prudent to use TIV in patients who were vaccinated at least once in the past 2 years but LAIV against pandemic strains may be more protective than inactivated vaccines, because the population will probably lack preexisting immunity."

High-dose Vaccine

A high-dose vaccine (Fluzone High-Dose) 4x the strength of standard flu vaccine was approved by the FDA in late 2009. This vaccine is intended for people 65 and over, who typically have weakened immune response due to normal aging. The vaccine produces a greater immune response than standard vaccine, but it is not yet known whether it provides greater protection against flu. Study results are expected in 2012. CDC recommends the high-dose vaccine for people 65 and over but expresses no preference between it and standard vaccine.

Vaccination Recommendations

Various public health organisations, including the World Health Organisation, have recommended that yearly influenza vaccination be routinely offered to patients at risk of complications of influenza and those individuals who live with or care for high-risk individuals, including:

- The elderly (UK recommendation is those aged 65 or above)
- Patients with chronic lung diseases (asthma, COPD, etc.)
- Patients with chronic heart diseases (congenital heart disease, chronic heart failure, ischaemic heart disease)
- Patients with chronic liver diseases (including cirrhosis)
- Patients with chronic renal diseases (such as the nephrotic syndrome)

- Patients who are immunosuppressed (those with HIV or who are receiving drugs to suppress the immune system such as chemotherapy and long-term steroids) and their household contacts
- People who live together in large numbers in an environment where influenza can spread rapidly, such as prisons, nursing homes, schools, and dormitories
- Healthcare workers (both to prevent sickness and to prevent spread to patients)
- *Pregnant women:* However, a 2009 review concluded that there was insufficient evidence to recommend routine use of trivalent influenza vaccine during the first trimester of pregnancy.
- Children from ages six months to two years.

Both types of flu vaccines are contraindicated for those with severe allergies to egg proteins and people with a history of Guillain-Barré syndrome.

Public Health Law Research, an independent organisation, published in 2009 several evidence briefs summarising the research assessing the effect of specific laws and policy on public health.

There is sufficient evidence supporting the effectiveness of requiring vaccinations as a condition for attending child care facilities and schools. There is insufficient evidence to assess the effectiveness of requiring vaccinations as a condition for specified jobs as a means of reducing incidence of specific diseases among particularly vulnerable populations.

There is strong evidence supporting the effectiveness of standing orders which allow healthcare workers without prescription authority to administer vaccines under defined circumstances as a public health intervention aimed at increasing vaccination rates.

Side Effects

- Side effects of the inactivated/dead flu vaccine injection include:
- Mild soreness, redness, and swelling where the shot was given
- Fever
- Aches
- These problems usually begin soon after the injection, and last 1–2 days.
- Side effects of the activated/live/LAIV flu nasal spray vaccine:
- Some children and adolescents 2–17 years of age have reported:
- Runny nose, nasal congestion or cough

- Fever
- Headache and muscle aches
- Wheesing
- Abdominal pain or occasional vomiting or diarrhea
- Some adults 18–49 years of age have reported:
- Runny nose or nasal congestion
- Sore throat
- Cough, chills, tiredness/weakness
- Headache
- More severe, but very rare side effects include:
- Life-threatening allergic reaction.

Some injection-based flu vaccines intended for adults in the United States contain thiomersal (also known as thimerosal). Despite some controversy in the media, the World Health Organisation has concluded that there is no evidence of toxicity from thiomersal in vaccines and no reason on grounds of safety to change to more-expensive single-dose administration. Although Guillain-Barré syndrome had been feared as a complication of vaccination, the CDC states that most studies on modern influenza vaccines have seen no link with Guillain-Barré.

A review has concluded that the 2009 H1N1 ("swine flu") vaccine has a safety profile similar to that of seasonal vaccine. Although one review gives an incidence of about one case per million vaccinations, a large study in China, reported in the NEJM covering close to 100 million doses of vaccine against the 2009 H1N1 "swine" flu found only eleven cases of Guillain-Barre syndrome, (0.1%) total incidence in persons vaccinated, actually lower than the normal rate of the disease in China, and no other notable side effects; "The risk-benefit ratio, which is what vaccines and everything in medicine is about, is overwhelmingly in favour of vaccination."

Getting infected by influenza itself increases both the risk of death (up to 1 in 10,000) and increases the risk of developing Guillain-Barré syndrome to a much higher level than the highest level of suspected vaccine involvement (approx. 10 times higher by recent estimates).

Flu Vaccine Manufacturing

Flu vaccine is usually grown in fertilised chicken eggs. In February preceding each fall's flu season (in the Northern hemisphere), three strains of flu are selected and chicken eggs inoculated.

As of November 2007, both the conventional injection and the nasal spray are manufactured using chicken eggs. The European Union has also approved Optaflu, a vaccine produced by Novartis using vats of animal cells. This technique is expected to be more scalable and avoid problems with eggs, such as allergic reactions and incompatibility with strains that affect avians like chickens.

A DNA-based vaccination, which is hoped to be even faster to manufacture, is currently in clinical trials, but has not yet been proven safe and effective. Research continues into the idea of a "universal" influenza vaccine (but no vaccine candidates have been announced) which would not need to be tailored to work on particular strains, but would be effective against a broad variety of influenza viruses.

In a 2007 report, the current global capacity of approximately 826 million seasonal influenza vaccine doses (inactivated and live) was double the current production of 413 million doses. In an aggressive scenario of producing pandemic influenza vaccines by 2013, only 2.8 billion courses could be produced in a six-month time frame. If all high- and upper-middle-income countries sought vaccines for their entire populations in a pandemic, nearly 2 billion courses would be required. If China pursued this goal as well, more than 3 billion courses would be required to serve these populations. Vaccine research and development is ongoing to identify novel vaccine approaches that could produce much greater quantities of vaccine at a price that is affordable to the global population.

An effective method of vaccine generation that bypasses the need for eggs is the construction of "influenza virus-like particle (VLP)". VLP is a non-egg, non-mammalian cell culture-based vaccine, purified from the supernatants of Spodoptera frugiperda Sf9 insect cells following infection of baculovirus vectors encoding an expression cassette made up of only three influenza virus structural proteins, hemagglutinin (HA), neuraminidase (NA), and matrix (M1) VLPs elicit antibodies that recognise a broader panel of antigenically distinct viral isolates compared to other vaccines in the hemagglutination-inhibition (HAI) assay.

H5N1

Vaccines have been formulated against several of the avian H5N1 influenza varieties. Vaccination of poultry against the ongoing H5N1 epizootic is widespread in certain countries. Some vaccines also exist for use in humans, and others are in testing, but none have been made available to civilian populations, nor produced in quantities sufficient to protect more than a tiny fraction of the Earth's population in the event of an H5N1 pandemic.

- Three H5N1 vaccines for humans have been licensed as of June 2008:
- Sanofi Pasteur's vaccine approved by the United States in April 2007,
- GlaxoSmithKline's vaccine Pandemrix approved by the European Union in May 2008, and
- CSL Limited's vaccine approved by Australia in June 2008.
- All are produced in eggs and would require many months to be altered to a pandemic version.

H5N1 continually mutates, meaning vaccines based on current samples of avian H5N1 cannot be depended upon to work in the case of a future pandemic of H5N1. While there can be some cross-protection against related flu strains, the best protection would be from a vaccine specifically produced for any future pandemic flu virus strain. Dr. Daniel Lucey, co-director of the Biohazardous Threats and Emerging Diseases graduate program at Georgetown University, has made this point, "There is no H5N1 pandemic so there can be no pandemic vaccine." However, "pre-pandemic vaccines" have been created; are being refined and tested; and do have some promise both in furthering research and preparedness for the next pandemic. Vaccine manufacturing companies are being encouraged to increase capacity so that if a pandemic vaccine is needed, facilities will be available for rapid production of large amounts of a vaccine specific to a new pandemic strain.

Problems with H5N1 vaccine production include:

- Lack of overall production capacity
- Lack of surge production capacity (it is impractical to develop a system that depends on hundreds of millions of 11 day old specialised eggs on a standby basis)
- The pandemic H5N1 might be lethal to chickens.

Cell culture (cell-based) manufacturing technology can be applied to influenza vaccines as they are with most viral vaccines and thereby solve the problems associated with creating flu vaccines using chicken eggs as is currently done.:

Currently, influenza vaccine for the annual, seasonal influenza program comes from four manufacturers. However, only a single manufacturer produces the annual vaccine entirely within the U.S. Thus, if a pandemic occurred and existing U.S.-based influenza vaccine manufacturing capacity was completely diverted to producing a pandemic vaccine, supply would be severely limited. Moreover, because the annual influenza manufacturing process takes place during most of the year, the time and capacity to produce vaccine against potential pandemic viruses for a stockpile, while continuing annual influenza vaccine production, is limited. Since supply will be limited, it is critical for HHS to be able to direct vaccine distribution in accordance with predefined groups; HHS will ensure the building of capacity and will engage states in a discussion about the purchase and distribution of pandemic influenza vaccine.

Vaccine production capacity: The protective immune response generated by current influenza vaccines is largely based on viral hemagglutinin (HA) and neuraminidase (NA) antigens in the vaccine. As a consequence, the basis of influenza vaccine manufacturing is growing massive quantities of virus in order to have sufficient amounts of these protein antigens to stimulate immune responses. Influenza vaccines used in the United States and around world are manufactured by growing virus in fertilised hens' eggs, a commercial process that has been in place for decades. To achieve current vaccine production targets millions of 11-day old fertilised eggs must be available every day of production.

In the near term, further expansion of these systems will provide additional capacity for the U.S.-based production of both seasonal and pandemic vaccines, however, the surge capacity that will be needed for a pandemic response cannot be met by egg-based vaccine production alone, as it is impractical to develop a system that depends on hundreds of millions of 11-day old specialised eggs on a standby basis. In addition, because a pandemic could result from an avian influenza strain that is lethal to chickens, it is impossible to ensure that eggs will be available to produce vaccine when needed.

In contrast, cell culture manufacturing technology can be applied to influenza vaccines as they are with most viral vaccines (e.g., polio vaccine, measles-mumps-rubella vaccine, chickenpox vaccine). In this system, viruses are grown in closed systems such as bioreactors containing large numbers of cells in growth media rather than eggs. The surge capacity afforded by cell-based technology is insensitive to seasons and can be adjusted to vaccine demand, as capacity can be increased or decreased by the number of bioreactors or the volume used within a bioreactor. In addition to supporting basic research on cell-based influenza vaccine development, HHS is currently supporting a number of vaccine manufacturers in the advanced development of cell-based influenza vaccines with the goal of developing U.S.-licensed cell-based influenza vaccines produced in the United States. The US government has purchased from Sanofi Pasteur and Chiron Corporation several million doses of vaccine meant to be used in case of an influenza pandemic of H5N1 avian influenza and is conducting clinical trials with these vaccines. Researchers at the University of Pittsburgh have had success with a genetically engineered vaccine that took only a month to make and completely protected chickens from the highly pathogenic H5N1 virus.

According to the United States Department of Health & Human Services

In addition to supporting basic research on cell-based influenza vaccine development, HHS is currently supporting a number of vaccine manufacturers in the advanced development of cell-based influenza vaccines with the goal of developing U.S.-licensed cell-based influenza vaccines produced in the United States. Dose-sparing technologies. Current U.S.-licensed vaccines stimulate an immune response based on the quantity of HA (hemagglutinin) antigen included in the dose. Methods to stimulate a strong immune response using less HA antigen are being studied in H5N1 and H9N2 vaccine trials. These include changing the mode of delivery from intramuscular to intradermal and the addition of immune-enhancing adjuvant to the vaccine formulation. Additionally, HHS is soliciting contract proposals from manufacturers of vaccines, adjuvants, and medical devices for the development and licensure of influenza vaccines that will provide dose-sparing alternative strategies.

Chiron Corporation is now recertified and under contract with the National Institutes of Health to produce 8,000–10,000 investi-gational doses of Avian Flu (H5N1) vaccine. MedImmune and Aventis Pasteur are under similar contracts.

The United States government hopes to obtain enough vaccine in 2006 to treat 4 million people. However, it is unclear whether this vaccine would be effective against a hypothetical mutated strain that would be easily transmitted through human populations, and the shelflife of stockpiled doses has yet to be determined. The *New England Journal of Medicine* reported on March 30, 2006 on one of dozens of vaccine studies currently being conducted. The Treanor et al. study was on vaccine produced from the human isolate (A/Vietnam/1203/2004 H5N1) of a virulent clade 1 influenza A (H5N1) virus with the use of a plasmid rescue system, with only the hemagglutinin and neuraminidase genes expressed and administered without adjuvant.

"The rest of the genes were derived from an avirulent egg-adapted influenza A/PR/8/34 strain.

The hemagglutinin gene was further modified to replace six basic amino acids associated with high pathogenicity in birds at the cleavage site between hemagglutinin 1 and hemagglutinin 2. Immunogenicity was assessed by microneu-tralisation and hemagglutination-inhibition assays with the use of the vaccine virus, although a subgroup of samples were tested with the use of the wild-type influenza A/Vietnam/1203/2004 (H5N1) virus."

The results of this study combined with others scheduled to be completed by spring 2007 is hoped will provide a highly immunogenic vaccine that is cross-protective against heterologous influenza strains.

On August 18, 2006. the World Health Organisation changed the H5N1 strains recommended for candidate vaccines for the first time since 2004. "The WHO's new prototype strains, prepared by reverse genetics, include three new H5N1 subclades.

The hemagglutinin sequences of most of the H5N1 avian influenza viruses circulating in the past few years fall into two genetic groups, or clades. Clade 1 includes human and bird isolates from Vietnam, Thailand, and Cambodia and bird isolates from Laos and Malaysia. Clade 2 viruses were first identified in bird isolates from China, Indonesia, Japan, and South Korea before spreading westward to the Middle East, Europe, and Africa.

The clade 2 viruses have been primarily responsible for human H5N1 infections that have occurred during late 2005 and 2006, according to WHO. Genetic analysis has identified six subclades of clade 2, three of which have a distinct geographic distribution and have been implicated in human infections:

- Subclade 1, Indonesia
- Subclade 2, Middle East, Europe, and Africa
- Subclade 3, China

On the basis of the three subclades, the WHO is offering companies and other groups that are interested in pandemic vaccine development these three new prototype strains:

- An A/Indonesia/2/2005-like virus
- An A/Bar headed goose/Quinghai/1A/2005-like virus
- An A/Anhui/1/2005-like virus.

Until now, researchers have been working on prepandemic vaccines for H5N1 viruses in clade 1. In March, the first clinical trial of a U.S.

vaccine for H5N1 showed modest results. In May, French researchers showed somewhat better results in a clinical trial of an H5N1 vaccine that included an adjuvant. Vaccine experts aren't sure if a vaccine effective against known H5N1 viral strains would be effective against future strains. Although the new viruses will now be available for vaccine research, WHO said clinical trials using the clade 1 viruses should continue as an essential step in pandemic preparedness, because the trials yield useful information on priming, cross-reactivity, and cross-protection by vaccine viruses from different clades and subclades."

As of November 2006, the United States Department of Health and Human Services still had enough H5N1 pre-pandemic vaccine to treat about 3 million people (5.9 million full-potency doses) in spite of 0.2 million doses used for research and 1.4 million doses that have begun to lose potency (from the original 7.5 million full-potency doses purchased from Sanofi Pasteur and Chiron Corp.).

The expected shelf life of seasonal flu vaccine is about a year so the fact that most of the H5N1 pre-pandemic stockpile is still good after about 2 years is considered encouraging.

Annual Reformulation of Flu Vaccine

Each year, three strains are chosen for selection in that year's flu vaccination by the WHO Global Influenza Surveillance Network. The chosen strains are the H1N1, H3N2, and Type-B strains thought most likely to cause significant human suffering in the coming season.

Due to the high mutation rate of the virus a particular vaccine formulation is effective for at most about a year. The World Health Organisation coordinates the contents of the vaccine each year to contain the most likely strains of the virus to attack the next year.

"The WHO Global Influenza Surveillance Network was established in 1952. The network comprises 4 WHO Collaborating Centres (WHO CCs) and 112 institutions in 83 countries, which are recognised by WHO as WHO National Influenza Centres (NICs). These NICs collect specimens in their country, perform primary virus isolation and preliminary antigenic characterisation.

They ship newly isolated strains to WHO CCs for high level antigenic and genetic analysis, the result of which forms the basis for WHO recommendations on the composition of influenza vaccine for the Northern and Southern Hemisphere each year."

The Global Influenza Surveillance Network's selection of viruses for the vaccine manufacturing process is based on its best estimate of which strains will be predominant the next year, amounting in the end to well-informed but fallible guesswork.

Formal WHO recommendations first issued in 1973; beginning 1999 there have been two recommendations per year, one for the northern hemisphere (N) and the other for the southern hemisphere (S).

History of past WHO seasonal influenza vaccine composition recommendations:

2002–2003 Northern Hemisphere Winter Season

The vaccines produced for the 2002–2003 season use:

- An A/New Caledonia/20/1999-like(H1N1);
- An A/Moscow/10/1999-like(H3N2);
- A B/Hong Kong/330/2001-like viruses.

2003 Southern Hemisphere Winter Season

The composition of influenza virus vaccines for use in the 2003 Southern Hemisphere influenza season recommended by the World Health Organisation was:

- An A/New Caledonia/20/99(H1N1)-like virus
- An A/Moscow/10/99(H3N2)-like virus (The widely used vaccine strain is A/Panama/2007/99)
- A B/Hong Kong/330/2001-like virus (Some vaccine strains used were B/Shandong/7/97, B/Hong Kong/330/2001, B/Hong Kong/1434/2002).

2003–2004 Northern Hemisphere Winter Season

The production of flu vaccine requires a lead time of about six months before the season. It is possible that by flu season a strain becomes common for which the vaccine does not provide protection. In the 2003–2004 season the vaccine was produced to protect against A/Panama, A/New Caledonia, and B/Hong Kong. A new strain, A/Fujian, was discovered after production of the vaccine started and vaccination gave only partial protection against this strain.

Nature magazine reported that the Influenza Genome Sequencing Project, using phylogenetic analysis of 156 H3N2 genomes, "explains the appearance, during the 2003–2004 season, of the 'Fujian/411/2002'-like strain, for which the existing vaccine had limited effectiveness" as due to an epidemiologically significant reassortment.

"Through a reassortment event, a minor clade provided the haemagglutinin gene that later became part of the dominant strain after the 2002–2003 season. Two of our samples, A/New York/269/2003 (H3N2) and A/New York/32/2003 (H3N2), show that this minor clade continued to circulate in the 2003–2004 season, when most other isolates were reassortants."

According to the CDC

During the 2003–2004 influenza season, influenza A (H1), A (H3N2), and B viruses co-circulated worldwide, and influenza A (H3N2) viruses predominated. Several Asian countries reported widespread outbreaks of avian influenza A (H5N1) among poultry. In Vietnam and Thailand, these outbreaks were associated with severe illnesses and deaths among humans.

In the United States, the 2003–2004 influenza season began earlier than most seasons, peaked in December, was moderately severe in terms of its impact on mortality, and was associated predominantly with influenza A (H3N2) viruses.

During September 28, 2003 – May 22, 2004, WHO and NREVSS collaborating laboratories in the United States tested 130,577 respiratory specimens for influenza viruses; 24,649 (18.9%) were positive. Of these, 24,393 (99.0%) were influenza A viruses, and 249 (1.0%) were influenza B viruses.

Among the influenza A viruses, 7,191 (29.5%) were subtyped; 7,189 (99.9%) were influenza A (H3N2) viruses, and two (0.1%) were influenza A (H1) viruses. The proportion of specimens testing positive for influenza first increased to >10% during the week ending October 25, 2003 (week 43), peaked at 35.2% during the week ending November 29 (week 48), and declined to <10% during the week ending January 17, 2004 (week 2).

The peak percentage of specimens testing positive for influenza during the previous four seasons had ranged from 23% to 31% and peaked during late December to late February.

As of June 15, 2004, CDC had antigenically characterised 1,024 influenza viruses collected by U.S. laboratories since October 1, 2003: 949 influenza A (H3N2) viruses, three influenza A (H1) viruses, one influenza A (H7N2) virus, and 71 influenza B viruses. Of the 949 influenza A (H3N2) isolates characterised, 106 (11.2%) were similar antigenically to the vaccine strain A/Panama/2007/99 (H3N2), and 843 (88.8%) were similar to the drift variant, A/Fujian/411/2002 (H3N2). Of the three A (H1) isolates that were characterised, two were H1N1 viruses, and one was an H1N2 virus. The hemagglutinin proteins of the influenza A (H1) viruses were similar antigenically to the hemagglutinin of the vaccine strain A/New Caledonia/20/99. Of the 71 influenza B isolates that were characterised, 66 (93%) belonged to the B/Yamagata/16/88 lineage and were similar antigenically to B/Sichuan/379/99, and five (7%) belonged to the B/Victoria/2/87 lineage and were similar antigenically to the corresponding vaccine strain B/Hong Kong/330/2001.

H9N2

In December 2003, one confirmed case of avian influenza A (H9N2) virus infection was reported in a child aged five years in Hong Kong.

The child had fever, cough, and nasal discharge in late November, was hospitalised for two days, and fully recovered. The source of this child's H9N2 infection is unknown.

H5N1

During January–March 2004, a total of 34 confirmed human cases of avian influenza A (H5N1) virus infection were reported in Vietnam and Thailand. The cases were associated with severe respiratory illness requiring hospitalisation and a case-fatality proportion of 68% (Vietnam: 22 cases, 15 deaths; Thailand: 12 cases, eight deaths). A substantial proportion of the cases were among children and young adults (i.e., persons aged 5–24 years). These cases were associated with widespread outbreaks of highly pathogenic H5N1 influenza among domestic poultry.

H7N3

During March 2004, health authorities in Canada reported two confirmed cases of avian influenza A (H7N3) virus infection in poultry workers who were involved in culling of poultry during outbreaks of highly pathogenic H7N3 on farms in the Fraser River Valley, British Columbia. One patient had unilateral conjunctivitis and nasal discharge, and the other had unilateral conjunctivitis and headache. Both illnesses resolved without hospitalisation.

H7N2

During the 2003–2004 influenza season, a case of avian influenza A (H7N2) virus infection was detected in an adult male from New York, who was hospitalised for upper and lower respiratory tract illness in November 2003.

Influenza A (H7N2) virus was isolated from a respiratory specimen from the patient, whose acute symptoms resolved.

The source of this person's infection is unknown.

2004 Southern Hemisphere Winter Season

The composition of influenza virus vaccines for use in the 2004 Southern Hemisphere influenza season recommended by the World Health Organisation was:

- An A/New Caledonia/20/99(H1N1)-like virus
- An A/Fujian/411/2002(H3N2)-like virus (A/Kumamoto/102/2002 and A/Wyoming/3/2003 were egg-grown A/Fujian/411/2002-like viruses)
- a B/Hong Kong/330/2001-like virus (B/Shandong/7/97, B/Hong Kong/330/2001 and B/Hong Kong/1434/2002 were among those used at the time. B/Brisbane/32/2002 was also available.)

2004–2005 Northern Hemisphere Winter Season

On the basis of antigenic analyses of recently isolated influenza viruses, epidemiologic data, and postvaccination serologic studies in humans, the Food and Drug Administration's Vaccines and Related Biological Products Advisory Committee (VRBPAC) recommended that the 2004–05 trivalent influenza vaccine for the United States contain A/New Caledonia/20/99-like (H1N1), A/Fujian/411/2002-like (H3N2), and B/Shanghai/361/2002-like viruses.

Because of the growth properties of the A/Wyoming/3/2003 and B/Jiangsu/10/2003 viruses, U.S. vaccine manu-facturers are using these antigenically equivalent strains in the vaccine as the H3N2 and B components, respectively. The A/New Caledonia/20/99 virus will be retained as the H1N1 component of the vaccine.

2005 Southern Hemisphere Winter Season

The composition of influenza virus vaccines for use in the 2005 Southern Hemisphere influenza season recommended by the World Health Organisation was:

- An A/New Caledonia/20/99(H1N1)-like virus;
- An A/Wellington/1/2004(H3N2)-like virus;
- A B/Shanghai/361/2002-like virus (B/Shanghai/361/2002, B/Jilin/20/2003 and B/Jiangsu/10/2003 were used at the time)

2005–2006 Northern Hemisphere Winter Season

The vaccines produced for the 2005–2006 season use:

- An A/New Caledonia/20/1999-like(H1N1);
- An A/California/7/2004-like(H3N2) (or the antigenically equivalent strain A/New York/55/2004);
- A B/Jiangsu/10/2003-like viruses.

In people in the United States, overall flu and pneumonia deaths were below those of a typical flu season with 84% Influenzavirus A and the rest Influenzavirus B. Of the patients who had Type A viruses, 80% had viruses identical or similar to the A bugs in the vaccine. 70% of the people testing positive for a B virus had Type B Victoria, a version not found in the vaccine.

"During the 2005–06 season, influenza A (H3N2) viruses predominated overall, but late in the season influenza B viruses were more frequently isolated than influenza A viruses. Influenza A (H1N1) viruses circulated at low levels throughout the season. Nationally, activity was low from October through early January, increased during February, and peaked in early March. Peak activity was less intense, but activity remained elevated for a longer period of time this season compared to the previous three seasons. The longer period of elevated activity may be due in part to regional differences in the timing of peak activity and intensity of influenza B activity later in the season."

2006 Southern Hemisphere Winter Season

The composition of influenza virus vaccines for use in the 2006 Southern Hemisphere influenza season recommended by the World Health Organisation was:

- An A/New Caledonia/20/99(H1N1)-like virus;
- An A/California/7/2004(H3N2)-like virus (A/New York/55/2004 was used at the time);
- A B/Malaysia/2506/2004-like virus.

2006–2007 Northern Hemisphere Winter Season

The 2006–2007 influenza vaccine composition recommended by the World Health Organisation on February 15, 2006 and the U.S. FDA's Vaccines and Related Biological Products Advisory Committee (VRBPAC) on February 17, 2006 use:

- An A/New Caledonia/20/99 (H1N1)-like virus;
- An A/Wisconsin/67/2005 (H3N2)-like virus (A/Wisconsin/67/2005 and A/Hiroshima/52/2005 strains);
- A B/Malaysia/2506/2004-like virus from B/Malaysia/2506/2004 and B/Ohio/1/2005 strains which are of B/Victoria/2/87 lineage.

2007 Southern Hemisphere Winter Season

The composition of influenza virus vaccines for use in the 2007 Southern Hemisphere influenza season recommended by the World Health Organisation on September 20, 2006 was:

- An A/New Caledonia/20/99(H1N1)-like virus,
- An A/Wisconsin/67/2005(H3N2)-like virus (A/Wisconsin/67/2005 and A/Hiroshima/52/2005 were used at the time),
- A B/Malaysia/2506/2004-like virus.

2007–2008 Northern Hemisphere Winter Season

The composition of influenza virus vaccines for use in the 2007–2008 Northern Hemisphere influenza season recommended by the World Health Organisation on February 14, 2007 was:

- An A/Solomon Islands/3/2006 (H1N1)-like virus;
- An A/Wisconsin/67/2005 (H3N2)-like virus (A/Wisconsin/67/2005 (H3N2) and A/Hiroshima/52/2005 were used at the time);
- A B/Malaysia/2506/2004-like virus.

In the US, the CDC reported in Feb 2008 that the H1N1 component was a good match (96%) to the infections occurring. But 87% of the H3N2 are A/Brisbane/10/2007-like viruses, which are a recent antigenic variant of the vaccine strain, A/Wisconsin. And 93% of the B viruses are in a B/Yamagata lineage that is relatively distinct from the vaccine strain B/Victoria lineage.

Only one of the three components was a good match; A/Wisconsin is moderately protective against the drifted A/Brisbane strain. 4.5% of those viruses tested are resistant to Oseltamivir, or Tamiflu—a significant increase over previous years. This vaccine has been described as 40% effective compared to other years that have been 85–95% effective.

2008 Southern Hemisphere Winter Season

The composition of virus vaccines for use in the 2008 Southern Hemisphere influenza season recommended by the World Health Organisation on September 17–19, 2007 was:

- An A/Solomon Islands/3/2006 (H1N1)-like virus;
- An A/Brisbane/10/2007 (H3N2)-like virus;
- A B/Florida/4/2006-like virus

2008-2009 Northern Hemisphere Winter Season

The composition of virus vaccines for use in the 2008-2009 Northern Hemisphere influenza season recommended by the World Health Organisation on February 14, 2008 was:

- An A/Brisbane/59/2007 (H1N1)-like virus;
- An A/Brisbane/10/2007 (H3N2)-like virus;

- a B/Florida/4/2006-like virus (B/Florida/4/2006 and B/Brisbane/3/2007 (a B/Florida/4/2006-like virus) were used at the time).

As of May 30, 2009: "CDC has antigenically characterised 1,567 seasonal human influenza viruses [947 influenza A (H1), 162 influenza A (H3) and 458 influenza B viruses] collected by U.S. laboratories since October 1, 2008, and 84 novel influenza A (H1N1) viruses. All 947 influenza seasonal A (H1) viruses are related to the influenza A (H1N1) component of the 2008-09 influenza vaccine (A/Brisbane/59/2007). All 162 influenza A (H3N2) viruses are related to the A (H3N2) vaccine component (A/Brisbane/10/2007).

All 84 novel influenza A (H1N1) viruses are related to the A/California/07/2009 (H1N1) reference virus selected by WHO as a potential candidate for novel influenza A (H1N1) vaccine. Influenza B viruses currently circulating can be divided into two distinct lineages represented by the B/Yamagata/16/88 and B/Victoria/02/87 viruses.

Sixty-one influenza B viruses tested belong to the B/Yamagata lineage and are related to the vaccine strain (B/Florida/04/2006). The remaining 397 viruses belong to the B/Victoria lineage and are not related to the vaccine strain."

2009 Southern Hemisphere Winter Season

The composition of virus vaccines for use in the 2009 Southern Hemisphere influenza season recommended by the World Health Organisation on September 17–19, 2008 was:

- An A/Brisbane/59/2007 (H1N1)-like virus;
- An A/Brisbane/10/2007 (H3N2)-like virus;
- A B/Florida/4/2006-like virus.

2009-2010 Northern Hemisphere Winter Season

The composition of virus vaccines for use in the 2009-2010 Northern Hemisphere influenza season recommended by the World Health Organisation on February 12, 2009 was:

- An A/Brisbane/59/2007 (H1N1)-like virus;
- An A/Brisbane/10/2007 (H3N2)-like virus;
- A B/Brisbane/60/2008-like virus.

Since the A/Brisbane/59/2007 (H1N1)-like virus used in the vaccine is an unrelated seasonal strain of influenza, it probably cannot create immunity to the new, non-seasonal strain of influenza A virus subtype H1N1 responsible for the 2009 swine flu outbreak.

2010 Southern Hemisphere Winter Season

- An A/California/7/2009 (H1N1)-like virus;
- An A/Perth/16/2009 (H3N2)-like virus;
- A B/Brisbane/60/2008-like virus.

The H1N1 strain used in this composition is the same strain used in the 2009 flu pandemic vaccine.

2010-2011 Northern Hemisphere Winter Season

The composition of virus vaccines for use in the 2010-2011 Northern Hemisphere influenza season recommended by the World Health Organisation on February 18, 2010 was:

- An A/California/7/2009 (H1N1)-like virus;
- An A/Perth/16/2009 (H3N2)-like virus;
- A B/Brisbane/60/2008-like virus.

The H1N1 strain used in this composition is the same strain used in the 2009 flu pandemic vaccine.

2011 Southern Hemisphere Winter Season

The composition of virus vaccines for use in the 2011 Southern Hemisphere influenza season recommended by the World Health Organisation on September 29, 2010 was:

- An A/California/7/2009 (H1N1)-like virus;
- An A/Perth/16/2009 (H3N2)-like virus;
- A B/Brisbane/60/2008-like virus.

The H1N1 strain used in this composition is the same strain used in the 2009 flu pandemic vaccine.

2011-2012 Northern Hemisphere Winter Season

The composition of virus vaccines for use in the 2011-2012 Northern Hemisphere influenza season recommended by the World Health Organisation on February 17, 2011 was:

- An A/California/7/2009 (H1N1)-like virus;
- An A/Perth/16/2009 (H3N2)-like virus;
- A B/Brisbane/60/2008-like virus.

The H1N1 strain used in this composition is the same strain used in the 2009 flu pandemic vaccine.

Flu Vaccine for Nonhumans

"Vaccination in the veterinary world pursues four goals: (i) protection from clinical disease, (ii) protection from infection with virulent virus,

(iii) protection from virus excretion, and (iv) serological differentiation of infected from vaccinated animals (so-called DIVA principle). In the field of influenza vaccination, neither commercially available nor experimentally tested vaccines have been shown so far to fulfil all of these requirements."

Horses

Horses with horse flu can run a fever, have a dry hacking cough, have a runny nose, and become depressed and reluctant to eat or drink for several days but usually recover in two to three weeks. "Vaccination schedules generally require a primary course of 2 doses, 3–6 weeks apart, followed by boosters at 6–12 month intervals. It is generally recognised that in many cases such schedules may not maintain protective levels of antibody and more frequent administration is advised in high-risk situations."

It is a common requirement at shows in the United Kingdom that horses be vaccinated against equine flu and a vaccination card must be produced; the FEI requires vaccination every six months.

Poultry

Poultry vaccines for bird flu are made on the cheap and are not filtered and purified like human vaccines to remove bits of bacteria or other viruses. They usually contain whole virus, not just hemagglutinin as in most human flu vaccines. Purification to standards needed for humans is far more expensive than the original creation of the unpurified vaccine from eggs. There is no market for veterinary vaccines that are that expensive. Another difference between human and poultry vaccines is that poultry vaccines are adjuvated with mineral oil, which induces a strong immune reaction but can cause inflammation and abscesses. "Chicken vaccinators who have accidentally jabbed themselves have developed painful swollen fingers or even lost thumbs, doctors said. Effectiveness may also be limited. Chicken vaccines are often only vaguely similar to circulating flu strains — some contain an H5N2 strain isolated in Mexico years ago. 'With a chicken, if you use a vaccine that's only 85 percent related, you'll get protection,' Dr. Cardona said. 'In humans, you can get a single point mutation, and a vaccine that's 99.99 percent related won't protect you.' And they are weaker than human vaccines. 'Chickens are smaller and you only need to protect them for six weeks, because that's how long they live till you eat them,' said Dr. John J.

Treanor, a vaccine expert at the University of Rochester. Human seasonal flu vaccines contain about 45 micrograms of antigen, while an experimental A(H5N1) vaccine contains 180. Chicken vaccines may contain less than 1 microgram. 'You have to be careful about extrapolating data from poultry to humans,' warned Dr. David E. Swayne, director of the agriculture department's Southeast Poultry Research Laboratory. 'Birds are more closely related to dinosaurs.'"

Researchers, led by Nicholas Savill of the University of Edinburgh in Scotland, used mathematical models to simulate the spread of H5N1 and concluded that "at least 95 per cent of birds need to be protected to prevent the virus spreading silently. In practice, it is difficult to protect more than 90 per cent of a flock; protection levels achieved by a vaccine are usually much lower than this."

The Food and Agriculture Organisation of the United Nations has issued recom-mendations on the prevention and control of avian influenza in poultry, including the use of vaccination. A filtered and purified Influenza A vaccine for humans is being developed and many countries have recom-mended it be stockpiled so if an Avian influenza pandemic starts jumping to humans, the vaccine can quickly be administered to avoid loss of life. Avian influenza is sometimes called avian flu, and commonly bird flu.

Pigs

Swine origin influenza virus (SoIV) vaccines are extensively used in the swine industry in Europe and North America. Most swine flu vaccine manufacturers include an H1N1 and an H3N2 SoIV strains. Swine influenza has become a greater problem in recent decades. Evolution of the virus has resulted in inconsistent responses to traditional vaccines. Standard commercial swine origin flu vaccines are effective in controlling the problem when the virus strains match enough to have significant cross-protection and custom (autogenous) vaccines made from the specific viruses isolated are created and used in the more difficult cases. SoIV vaccine manufacture Novartis paints this picture: "A strain of swine origin influenza virus (SoIV) called H3N2, first identified in the US in 1998, has brought exasperating production losses to swine producers. Abortion storms are a common sign. Sows go off feed for two or three days and run a fever up to 106°F. Mortality in a naïve herd can run as high as 15%."

Dogs

In 2004, Influenza A virus subtype H3N8 was discovered to cause canine influenza. Because of the lack of previous exposure to this virus, dogs have no natural immunity to this virus. However a vaccine is now available.

Computer-assisted Vaccine Design

A new parametre has been defined to quantify the antigenic distance between two H3N2 influenza strains. This parametre was used to measure antigenic distance between circulating H3N2 strains and the closest vaccine component of the influenza vaccine. For the data between 1971 and 2004, the measure of antigenic distance correlated better with efficacy in humans of the H3N2 influenza A annual vaccine than did current measures of antigenic distance such as phylogenetic sequence analysis or ferret antisera inhibition assays. This measure of antigenic distance could be used to guide the design of the annual flu vaccine. The antigenic distance combined with a multiple-strain avian influenza transmission model was used to study the threat of simultaneous introduction of multiple avian influenza strains. Population at Risk (PaR) can be used to quantify the risk of a flu pandemic and to calculate the improvement that a multiple vaccine offers.

Chapter 11

Disease Prevention Management

Various aspects of poultry health maintenance and disease prevention are outlined. The present day health problems are often multi factorial and require special attention to avoid disastrous consequences. Health is a state of complete physical, mental and social well being of poultry, which is an important factor determining the profitability of poultry production. Disease represents a restraint to reproductive potential and yield, feed conversion efficiency and product quality. The causative factors of disease are classified as agent, host and environment. The interaction of these three factors is required to start disease in birds. Intensive health monitoring of elite primary breeders, GGPs, GPs is essential for effective prevention of vertically transmitted disea ses. Integrated disease prevention and control involves all the measures designed to prevent or reduce the incidence, prevalence and consequences of the disease. Broadly these measures include 1. Controlling the reservoir or source of infection. 2. Breaking the routes of transmission and 3. Manipulating the susceptible host. The practical use of vaccines to prevent or reduce disease losses in poultry is discussed. The objectives of vaccination, the types of vaccine, vaccination programme, combined vaccines and vaccination failures are presented in a non technical manner.

The concept of preventive medicine in poultry is now more important than ever, because the occurrence of disease will limit the performance of the bird and prevent it from achieving its maximum genetic potential. The present day poultry health problems are seldom unifactorial, on the contrary. They are very often multi factorial, requiring special attend on to avert disastrous consequences. These problems, which may not be frank disease hut results in sub and time consuming to overcome. It is much better to avoid those happenings in the first place through an integrated system of health maintenance and disease prevention. Successful prevention depends on knowledge of causation, dynamics of transmission, identification of risk factors and flocks at risk, availability of prophylactic or early detection and treatment measures, application of these measures to appropriate or flock and continuous evaluation procedures applied.

Structure of Poultry Industry and Disease Control

The basic structure of the poultry in-dustry can be regarded as a Breeding Pyramid. The apex of the pyramid is occupied by Elite (Primary) Breeders, Great Grand Parents (GGPs) and Grand Parents (GPs), followed by Parents in the middle and commercials at the bottom of the pyramid (Hunton, 1993). The major genetic manipulations take place at elite and GGPs level. This involves careful recording of the main production parameters and also, selection for desired traints. At this level intensive health monitoring is essential, so that diseases which can be vertically transmitted or which can depress performance are excluded.

At this stage, the pathogens detrimental to the production and health of birds are to be eradicated from the elite breeders to GGPs, GPs and Parents can produce devastating results in the commercial poultry industry as hundreds or thousands of commercial birds are descended from each bird of the elite breeder flocks. So the disease status, immune system and genotype of elite flocks have potentially profound effect on the health of commercial flocks. The need to protect elite and GGPs from catastrophic and vertically transmitted diseases dictates high levels of biosecurity.

Concept of Health and Disease

Health is one of the most important factors determining profitability of all systems of poultry production. It is defined as a state of being that is not merely absence of disease, but a state of complete physical, mental and social well being of poultry. Health and disease lie along a continuum and there is no single cut-off point. The lowest point on the heath disease spectrum is death and the highest point corresponds to perfect health. It is thus obvious that health fluctuates within a range of optimum well being to various levels of dysfunction, including the state of total dysfunction, i.e. death.

Health and Disease Spectrum:

1. Perfect Health
2. Better Health

3. Freedom from Disease
4. Unrecognized Disease
5. Mild Disease
6. Severe Disease
7. Death.

Health status not only influences the average performance of a flock but also its variability. The perfect health implies that a bird should be able to express as completely as possible the potentialities of its genetic heritage. This is possible only when a bird is allowed to live is healthy relationship with its environment an environment that transforms genetic potentialities into phenotypic realities.

The disease is Just the opposite of health i.e., any deviation from normal functioning or state of complete physical or mental wellbeing. The term "disease" literally means "without ease" (uneasiness) disease, the opposite of ease when something is wrong with body function. Disease represents a restraint to reproductive potential and yield, feed conversion efficiency and product quality.

The causative factors of disease may be classified as Agent (Pathogen), Host (Bird) and Environment (Jordan, 1996). These three factors are referred to epidemiological triad. The mere presence of agent, host and favourable environmental factors is not sufficient to start the disease in bird. What is required is an interaction of these three factors to initiate disease process in the bird. The agent, host and environment operating in combination, determines not only the onset of disease which may range from a single case to epidemics but also the distribution of disease in the flock.

Agent Factors:

1. *Biologic agents:* The biologic agent may include viruses, Chlamydia, rickettdiae, mycophasmas, bacteria, fungi and para-sites, These agents exhibit certain biological properties viz., a. Antigenic variation, b Variation En virulence, c. Variation in immunogenic, d. Infectivity (numbers necessary to establish infection in the host), e. Tropism, f. Survival within and outside the host and g. Resistance/Susceptibility to antimicrobial and disinfectants.
2. Nutritional agents
3. Physical agents
4. Chemical agents
5. Mechanicalagents.

Host Factors (Intrinsic) :

1. Age
2. Gender
3. Immune status
4. Stress (associated with point of lay or social stress).

Environment factors (Extrinsic):

1. Climate
   a) Season
   b) Ambient Temperature
   c) Humidity
2. Air Quality
   a) Dust
   b) Gases like NHs, C02. And CH4
   c) Free radicals.
3. Housing
   a) Direction
   b) Ventilation.
4. Lighting
   a) Intensity
   b) Colour
   c) Duration
5. Management
   a) Floor/Cage rearing
   b) All-in-all out
   c) Multiple ages
   d) Vaccination & Medication
   e) Debarking, Dubbing & Sexing
   f) Handling
   g) Transportation
   h) Hygiene and disinfection
   i) Disposal of dead birds, litter and hatchery waste.
6. Nutrition
   a) Deficiency or excess of nutrients
   b) Water quality and supply

c) Feed toxins
d) Feed contaminants
e) Change of diet
f) Drugs-antimicrobials, growth promoters

7. Biological
   a) Insects
   b) Rodents
   c) Wild birds
   d) Opportunistic infectious agents
8. Social
   a) Population density
   b) Sex ratio.

Disease Transmission

Infectious diseases do not occur by chance. It is required for a harmful infection environment. The disease causing organisms are introduced or transmitted to a poultry flock in a variety of ways.

- Carriers within a flock
- Birds recently acquired from an out side flock
- Eggs from infected breeder flocks
- Human feet, hands and clothes
- Dust, feathers and manure on equipment and supplies, such as trucks, coops and egg filler flats
- Pet birds, predators, rodents, flies and insects
- Live or contaminated vaccines
- Contaminated feed/feed ingredients
- Water
- Air.

Disease Prevention and Control

Disease control involves all the measures designed to prevent or reduce the incidence, prevalence and consequences of the disease. The basic approach in controlling disease is to identify the weak points of diseases and break the weakest links in the chain of transmission (Pattison, 1993). Broadly these measures include:

1) controlling the reservoir or source of infection,
2) breaking the routes.

Controlling the Reservoirs and Source of Infection

If the first link in the chain of causation (i.e. the disease agent) is deemed to be the weakest fink, logically, the most desirable control measure would be to eliminate the reservoir source. This includes all the measures directed to reduce the quantity of the agent available for dissemination.

1. Early identification of reservoir/source of infection
2. Treatment of affected birds
3. Epidemiological investigation
4. Isolation (separation) of infected birds from healthy birds
5. Serological testing and removal of carriers
6. Hygienic disposal of dead birds, litter and hatchery waste
7. through cleaning, washing and disinfection of farm houses, equipment
8. Feed and Water quality control
9. Hygienic hatching of chicks.

Breaking Transmission Routes

A major aspect of communicable disease control relates to "breaking the chain of transmission". Microbes travel from place to place by chicken crates, egg filler flats, trucks, other equipment and people. They can be transmitted from one flock to the other through workers. The organisms are found on the skin and in the digestive and respiratory organs of dogs, cats, rats, mice and wild free flying birds, Flies, beetles and other insects are also notorious spreaders of disease causing organisms. Blocking the routes of transmission imply on control of environmental factors, that is, to bring about an adjusted equilibrium between host and environment through encouraging some ecological influences and inhibiting others (Woodger.1996).

- Human traffic control, such as locking the doors and banning the visitors.
- Providing foot dips (at the entrance of all houses), wheel dips at the entrance of the farm
- Aerial and water sanitation
- Pest control (flies, mosquitoes, ants, rodents, litter beetles, lice, bed bugs, wild birds, cockroaches, spiders, mites and weeds)
- All-in-all out farm management
- Restriction of movement of personnel, feed, vehicles and farm equipment

- Personnel hygiene
- Avoiding of on farm trading.

Manipulating the Susceptible Host (Birds at Risk)

The third link in the chain of transmission is the susceptible host or birds at risk. One effective way of controlling the spread of infection is to strengthen the host defense (Cheville, 1979). Under certain circumstances this may be accomplished by the following ways:

1. Genetic improvement of disease resistance by conventional selection and genetic engineering (gene, transfer).
2. Manipulation of immune system by
   a) Vaccination
   b) Avoiding social, environmental and managerial stress.
   c) Providing good quality, well balanced feed
   d) Immunomodulation.

Vaccination

Over decades of research, avian immune system has been successfully manipulated to benefit the poultry industry. The efficient functioning of avian immune system is judged through successful vaccination programme.

It is one of the most powerful and cost effective weapons of poultry preventive medicine. Diseases with both demonstrated and predicted potential for producing disastrous losses have been controlled through vaccination.

Vaccination is the second line of defense against disease losses, and it keeps disease losses within acceptable limits when sanitary management fails. Vaccination has to be designed according to the needs of the situation (Beard, 1979).

Vaccination objectives: The first key to develop an effective vaccination programme is to determine the vaccination objectives. The most important objectives of vaccination in poultry are as follows:

1. To prevent or reduce disease losses due to morbidity or mortality in vaccinates.
2. To prevent spread of diseases
3. To provide parental antibodies to chicks of immunized hens
4. To prevent adverse effects of vertically transmitted diseases in chicks
5. To prevent diseases which infect young chicks and subsequently results into development of unprofitable adults?

6. To protect against disease which damage immune system?
7. To reduce condemnation at slaughter
8. To prevent or reduce losses due to poor egg quality.

Types of Vaccines: Once the objectives are established and decision is made to pursue the vaccination pro-gramme, the type of vaccine, frequency/timing and route of administration should be matched to those objectives. Vaccines for poultry are of two types, either live or killed, each having specific advantages or uses. Live vaccines contain a virus or bacteria that must infect the chicken and multi-ply in its body to produce immunity, hope fully with minimal reaction. Advantages of live vaccines are ease of administration, low price, rapid onset of immunity and a broader scope of protection, since chickens are exposed to all stages of the replication virus. Killed vaccine is prepared from bacteria or virus which have been inactivated and processed. Killed vaccines are usually combined with adjuvant such as aluminum hydroxide or oil. Adjuvant enhance the immune response by increasing the stability of vaccine in the body which stimulates the immune system for a longer period of time. Advantages of Rifled vaccines are assurance of administration of a uniform dose, safety, development of uniform levels of immunity and increased product stability.

Vaccination Programme: Vaccination programme vary with the type of poultry and severity of the field challenge. The immune responses needed in broilers are for a short duration because of the short life span of the bird (six weeks). Breeders and layers require different programme because of longer life span (Aini, 1980). When designing a vaccination programme, one must first determine the disease challenges faced by the poultry industry in the region. To be effective, a vaccination challenge should be provided before any natural exposure to infection.

Bibliography

Abbas, M.A.: *Effect of Organic Amendments and EM on Crop Production in Pakistan,* SP, Brazil. Pub. USDA. Washington, D.C., 19991.

Appleby, M.C & Hughes, B.O. & Elson, H.A.: *Poultry Production Systems, Behaviour, Management and Welfare,* CAB International, NY, 1992.

Berton, V. and Mudd, D.: *Profitable Poultry: Raising Birds on Pasture,* USDA's Sustainable Agriculture Network (SAN), Washington, DC, 2001.

Boorman, K.N.: *Poultry Feedstuffs, Supply Composition and Nutritive Value,* CABI Publishing, UK, 2002.

Chenoweth, H.: *Free-Range Poultry,* Free-Range Poultry Production and Marketing, Creola, Ohio, 2001.

Cotterill, O.J.: *Egg Science and Technology,* Food Products Press, Imprint of Haworth Press, New York, London, 1995.

Dekkers, J.C.M., H.H. Zhao, and RL. Fernando: *Linkage Disequilibrium Mapping in Livestock,* Belo Horizonte, Brazil, 2006.

Donald, E. Green: *A History of the Oklahoma State University Division of Agriculture,* Oklahoma State University, Stillwater, 1990.

Elson, H.A.: *Poultry Production Systems, Behaviour, Management and Welfare,* CAB International, NY, 1992.

Fanatico, A.: *Pastured Poultry: A Heifer Project International Case Study Booklet,* Little Rock, AR, 2000.

Ferguson, M.W.J.: *Egg Incubation, Its Effects on Embryonic Development in Birds and Reptiles,* Cambridge University Press, UK, 1991.

Gillespie, J.R.: *Modern Livestock and Poultry Production 7th edition,* Thomson Delmar Learning, 2004.

Grist, A.: *Poultry Inspection, Anatomy, Physiology and Disease Conditions,* Nottingham University Press, UK, 2006.

Hernandez, J.: *Optimum Egg Quality: A Practical Approach,* 5M Publishing, U.K, 2007.

Huopalahti, R. & Lopez-Fandino, R. & Anton, M. & Schade, R.: *Bioactive Egg Compounds,* Springer, Delhi, 2007.

J. Dennis, Lord: *Encyclopedia of Southern Culture*, University of North Carolina Press, Chapel Hill, 1989.

Jordan, F & Pattison, M & Alexander, D & Faragher, T.: *Poultry Diseases,* WB Saunders, UK, 2002.

Larbier, M.; Leclercq, B.: *Nutrition and Feeding of Poultry*, Nottingham University Press, U.K., 1994.

Leclercq, B.: *Nutrition and Feeding of Poultry*, Nottingham University Press, U.K., 1994.

Mandal, A.B. : *Nutrition and Disease Management of Poultry*, International Book Distributing Co, Delhi, 2004.

Mudd, D.: *Profitable Poultry: Raising Birds on Pasture*, USDA's Sustainable Agriculture Network (SAN), Washington, DC, 2001.

Nicholls, C.: *The Workboot Series: The Story of Eggs in Australia,* Kondinin Group, Cloverdale W.A., 2005.

Nicholls, C.: *The Workboot Series: The Story of Eggs,* Kondinin Group, Cloverdale W.A., 2005.

Panda, A K ; S V Rama Rao and M R Reddy: *Growth Promoters in Poultry : Novel Concepts*, International Book Dist, Delhi, 2008.

Prasad, Sushil : *Handbook of Poultry Production : A Practical Guide*, Enkay Publishing House, Delhi, 2011.

Rakesh Kumar Shukla *: Handbook of Poultry Diseases : A Bedside-Guide*, International Book Dist, Delhi, 2006.

Roberts, J.R. & Ball, W.: *Egg Quality Guidelines for the Australian Egg Industry,* AECL Publication, UK, 2004.

Saxena, H.C. : *Poultry Feed Technology : Feed Formulation and Manufacturing*, International Book, Delhi, 2006.

Shukla, Rakesh Kumar *: Handbook of Poultry Diseases : A Bedside-Guide*, International Book Dist, Delhi, 2006.

Tullett, S.G.: *Poultry Science Symposium Number 22*. Butterworth-Heinemann, NJ, 1991.

Vegad, J L *: Poultry Diseases : A Guide for Farmers and Poultry Professionals*, International Book Dist, Delhi, 2008.

Verma, S. S. . : *Microbiological Changes on Chicken Carcasses During Processing*, Indian Journal of Poultry Science, 1989.

Whitehead, C.C.: *Bone Biology and Skeletal Disorders in Poultry,* Carfax Publishing Company, U.K., 1992.

Wiseman, J. & Garnsworthy, P.C.: *Recent Developments in Poultry Nutrition,* University Press, India, 1999.

Yamamoto, T. & Juneja, L.R. & Hatta, H.: *Hen Eggs: Basic and Applied Science,* CRC Press, Delhi, 1996.

Index

❑❑❑